National Security Space Defense and Protection

Public Report

Committee on National Security Space Defense and Protection

Division on Engineering and Physical Sciences

A Report of

The National Academies of
SCIENCES · ENGINEERING · MEDICINE

THE NATIONAL ACADEMIES PRESS
Washington, DC
www.nap.edu

THE NATIONAL ACADEMIES PRESS **500 Fifth Street, NW** **Washington, DC 20001**

This activity was supported by Contract 2014-14041100003-0004 between the Office of the Director of National Intelligence and the National Academy of Sciences. Any opinions, findings, or conclusions expressed in this publication do not necessarily reflect the view of any organization or agency that provided support for the project.

International Standard Book Number-13: 978-0-309-44748-5
International Standard Book Number-10: 0-309-44748-8
Digital Object Identifier: 10.17226/23594

Limited copies of this report are available from:

Air Force Studies Board
National Research Council
500 Fifth Street, NW
Washington, DC 20001
(202) 334-3111

Additional copies are available from:

The National Academies Press
500 Fifth Street, NW
Keck 360
Washington, DC 20001
(800) 624-6242 or (202) 334-3313
http://www.nap.edu

Printed in the United States of America

Suggested citation: National Academies of Sciences, Engineering, and Medicine. 2016. *National Security Space Defense and Protection: Public Report*. Washington, DC: The National Academies Press. doi: 10.17226/23594.

The National Academies of

SCIENCES · ENGINEERING · MEDICINE

The **National Academy of Sciences** was established in 1863 by an Act of Congress, signed by President Lincoln, as a private, nongovernmental institution to advise the nation on issues related to science and technology. Members are elected by their peers for outstanding contributions to research. Dr. Marcia McNutt is president.

The **National Academy of Engineering** was established in 1964 under the charter of the National Academy of Sciences to bring the practices of engineering to advising the nation. Members are elected by their peers for extraordinary contributions to engineering. Dr. C. D. Mote, Jr., is president.

The **National Academy of Medicine** (formerly the Institute of Medicine) was established in 1970 under the charter of the National Academy of Sciences to advise the nation on medical and health issues. Members are elected by their peers for distinguished contributions to medicine and health. Dr. Victor J. Dzau is president.

The three Academies work together as the **National Academies of Sciences, Engineering, and Medicine** to provide independent, objective analysis and advice to the nation and conduct other activities to solve complex problems and inform public policy decisions. The Academies also encourage education and research, recognize outstanding contributions to knowledge, and increase public understanding in matters of science, engineering, and medicine.

Learn more about the National Academies of Sciences, Engineering, and Medicine at **www.national-academies.org**.

The National Academies of

SCIENCES · ENGINEERING · MEDICINE

COMMITTEE ON NATIONAL SECURITY SPACE DEFENSE AND PROTECTION

JAMES O. ELLIS, JR., U.S. Navy (retired), Stanford University, *Co-Chair*
MARTIN C. FAGA, MITRE Corporation (retired), *Co-Chair*
ALLISON ASTORINO-COURTOIS, National Security Innovations, Inc.
OWEN C. BROWN, SAIC
VINCENT W.S. CHAN, Massachusetts Institute of Technology
MICHAEL D. GRIFFIN, Schafer Corporation
RAYMOND JEANLOZ, University of California at Berkeley
DAVID A. KOPLOW, Georgetown University
L. ROGER MASON, JR., Noblis
JOHN A. MONTGOMERY, Naval Research Laboratory
SCOTT PACE, George Washington University
THOMAS E. ROMESSER, Independent Consultant
WILLIAM L. SHELTON, U.S. Air Force (retired)
BOB THOMSON, Independent Consultant
DAVID M. VAN WIE, Johns Hopkins University Applied Physics Laboratory
DEBORAH L. WESTPHAL, Toffler Associates

Staff

JOAN FULLER, Board Director
ALAN H. SHAW, Deputy Board Director
CARTER W. FORD, Study Director
DIXIE GORDON, Information Officer
MARGUERITE E. SCHNEIDER, Administrative Coordinator

Preface

As part of the Fiscal Year 2014 National Defense Authorization Act, Congress directed the Office of the Director of National Intelligence (ODNI) and the Office of the Secretary of Defense (OSD) to contract with the National Research Council (NRC) to undertake a study on U.S. national security space defense and protection.[1] Somewhat at the same time, ODNI and OSD undertook a series of related initiatives, including the Space Strategic Portfolio Review (SPR), the congressionally directed Space Protection Strategy (SPS), and the Space Security and Defense Program (SSDP). In January 2015, the NRC approved the study terms of reference (TOR) and appointed a committee of experts to do the following:[2]

> 1. Review the range of options available to address threats to space systems, in terms of deterring hostile actions, defeating hostile actions, and surviving hostile actions.[3]

---

[1] For more information, see P.L. 113-66, December 26, 2013. Available at https://www.congress.gov/113/plaws/publ66/PLAW-113publ66.pdf. Accessed June 8, 2015.

[2] Appendix A provides biographies of the committee members. The committee includes experts with experience in academia, government, and industry, combined with many years in U.S. combatant commands and major commands, intelligence community, space law, spacecraft survivability, systems engineering, system architecting, space operations, space acquisition, cyberdefense, strategic deterrence, and high-altitude electromagnetic pulse.

[3] "System" is defined for purposes of this report as "a functionality, physically, and/or behaviorally related group of regularly interacting or interdependent elements; that group of elements forming a unified whole." "Space systems," in turn, are defined as "all of the devices and organizations forming the space network."

2. Assess potential strategies and plans to counter such threats, including resilience, reconstitution, disaggregation, and other appropriate concepts.

3. Assess existing and planned architectures, warfighter requirements, technology development, systems, workforce, or other factors related to addressing such threats.

4. Recommend architectures, capabilities, and courses of action to address such threats and actions to address affordability, technology risk, and other potential barriers or limiting factors in implementing such courses of action.

STUDY METHODOLOGY AND CAVEATS

The committee held eight meetings, beginning in February 2015 and ending in October 2015, to collect information and draft findings and recommendations.[4] With the understanding of Congress, OSD, and ODNI, the authoring committee produced two stand-alone classified reports to address the TOR and delivered them to the sponsors in August 2015 and December 2015, respectively. Collectively, the committee provided 30 findings and 18 recommendations to the sponsors. The requirement to report initial findings and recommendations to key stakeholders no later than August 15, 2015, essentially divided this study into two overlapping phases: phase one, February-August 2015, which addressed TOR items 1 and 2; and phase two, July-December 2015, which addresses TOR items 3 and 4. Report 2 contained analysis, findings, and recommendations that complemented those found in Report 1. The committee was granted rich access to documents and officials involved with intelligence collection, policy and planning, strategy, budgetary processes, and organizational realignments and assignments. In addition, the committee invited industry and federally funded research and development centers to participate at a 1-day session in April 2015, in conjunction with its third full committee meeting.

Importantly, no independent modeling or analysis was completed by the committee; rather, the information gathered from interviews, documents, and briefings, together with the expertise and experience of committee members, served as the bases for the committee's work. This unclassified summary, while admittedly brief due to government classification requirements, reflects the unclassified content of both classified reports. This unclassified summary is primarily a policy discussion. The reasons behind this focus are twofold. First, the system technologies themselves, the overall system architectures, and the operational aspects to their employment are predominantly classified at very high levels. Second, the committee observes that, as the summary states, there are major national policy issues that need to be addressed in order for the nation to formulate a wise and coherent approach to space defense and protection. On a macro level, two primary themes emerged from

[4] Appendix B provides a listing of invited speakers for both phases of the study.

this study regarding potential solutions to the threats facing U.S. space assets. First, the state of organization and coordination among various government activities is evolving and necessarily so. Second, there is an urgent need to create relevant national policies to guide the creation of responses to these threats; this includes educating the public so that it can understand and participate in potential solutions in whatever capacity makes sense.

ORGANIZATION OF THE REPORT

Chapter 1 provides an overall context for the report and explains how space is no longer a domain exclusively for national security. It discusses commercial trends at a high level that will help shape the future in space. Chapter 2 then describes measures for preserving national security space-enabled capabilities, including system protection measures, deterrence, and potential international avenues, such as regimes.

Acknowledgment of Reviewers

This report has been reviewed in draft form by individuals chosen for their diverse perspectives and technical expertise. The purpose of this independent review is to provide candid and critical comments that will assist the institution in making its published report as sound as possible and to ensure that the report meets institutional standards for objectivity, evidence, and responsiveness to the study charge. The review comments and draft manuscript remain confidential to protect the integrity of the deliberative process. We wish to thank the following individuals for their review of this report:

Rita R. Colwell, Johns Hopkins University and University of Maryland,
Gurudas Ganguli, Naval Research Laboratory,
Anita K. Jones, University of Virginia,
Paul G. Kaminski, Technovation, Inc.,
Donald A. Lewis, The Aerospace Corporation,
Lester L. Lyles, U.S. Air Force (retired),
Grant Stokes, MIT Lincoln Laboratory, and
Peter J. Weinberger, Google, Inc.

Although the reviewers listed above have provided many constructive comments and suggestions, they were not asked to endorse the conclusions or recommendations, nor did they see the final draft of the report before its release. The review of this report was overseen by John Stenbit, U.S. Department of Defense (retired), who was responsible for making certain that an independent examination of this report was carried out in accordance with institutional procedures and that all review comments were carefully considered. Responsibility for the final content of this report rests entirely with the authoring committee and the institution.

Contents

SUMMARY 1

1 CONTEXT FOR STUDY 8
Introduction, 8
Space-Enabled Capabilities Are Increasingly Shared, 10
The Accessibility of Space, 14
Domestic and International Consumer Markets, 16
Government and Commercial Sectors, 19
Consumer Demands Help Drive Innovation in Space, 22
The Vitality of Space, 23
National Security Uses of Space, 24
Low Earth Orbits and Functions, 24
Medium Earth Orbits and Functions, 24
Geosynchronous Earth Orbits and Functions, 25
Highly Elliptical Orbits and Functions, 25
Final Thoughts, 26

2 SELECTED ISSUES RELATED TO NATIONAL SECURITY SPACE DEFENSE AND PROTECTION 27
Introduction, 27
The Characterization of Space in National Discourse, 29

The Role of Space in National Security, 29
Space Services: Classifying What Is at Stake, 32
Threats to Space Systems and Services, 33
Defending and Protecting National Security Space Assets: Space Defense Triad, 34
System Protection Measures, 36
Deterrence Measures, 37
Credibility of a Deterrent Threat, 38
Capability of Responding, 39
Communicating Deterrence Messages, 40
Coalition Formation and International Regimes, 43
Final Thoughts, 44

APPENDIXES

A Biographical Sketches of Committee Members 47
B Meetings and Speakers 58

Acronyms

A2AD	antiaccess area denial
AJ	antijam
AoA	analysis of alternatives
ASAT	antisatellite
CAGR	compound annual growth rate
COMSAT	communications satellite
DoD	Department of Defense
EHF	extremely high frequency
EMP	electromagnetic pulse
GEO	geostationary/geosychronous (orbit)
GPS	Global Positioning System
IC	Intelligence Community
ISR	intelligence, surveillance, and reconnaissance
LEO	low Earth orbit
MILSATCOM	military satellite communications

NRC	National Research Council
NSS	national security space
ODNI	Office of the Director for National Intelligence
OPLAN	operational plan
OSD	Office of the Secretary of Defense
PNT	position, navigation, and timing
RF	radio frequency
SATCOM	satellite communications
SBIR	space-based infrared
SIGINT	signals intelligence
SPR	Space Strategic Portfolio Review
SPS	Space Protection Strategy
SSD	Space Security and Defense Program
TOR	terms of reference
TTP	tactic, technique, and procedure

Summary

> We set sail on this new sea because there is new knowledge to be gained, and new rights to be won, and they must be won and used for the progress of all people. For space science, like nuclear science and all technology, has no conscience of its own. Whether it will become a force for good or ill depends on man, and only if the United States occupies a position of pre-eminence can we help decide whether this new ocean will be a sea of peace or a new terrifying theater of war.
>
> I do not say that we should or will go unprotected against the hostile misuse of space any more than we go unprotected against the hostile use of land or sea, but I do say that space can be explored and mastered without feeding the fires of war, without repeating the mistakes that man has made in extending his writ around this globe of ours.
>
> President John F. Kennedy
> speech at Rice University
> September 12, 1962

The national security of the United States is inextricably linked to space and our unimpeded access to the capabilities resident in or traveling through that domain. Since the dawn of the Space Age, all those who have been a part of what was once a race between two superpowers and is now a $315 billion global enterprise, have implicitly understood this linkage. Over more than six decades, that reliance on space systems has deepened and broadened. What was once only a realm of exploration and national security has grown to include a commercial element that has become so ubiquitous that it has led us to fundamentally redefine the term national security space. President Kennedy was not the first to draw the analogy between space and the oceans of the world. The literature is sprinkled with references to space "ships," interplanetary "voyages," and star "fleets." Even the term "astronaut" is a combination of two Greek words, for "star" and "sailor." In many ways, the analogy is apt in that space exploration, initially, and exploitation, ultimately, have parallels in mankind's first tentative maritime endeavors. Sea-borne voyages of discovery led to the establishment of trade routes, colonial expansion, and, finally, contests for influence and security in the new domain.

The significant difference, of course, between the creation of global maritime policy and practice and that of the space domain is time. The technologies, customary behaviors, conventions and, eventually, treaties governing military and commercial naval activity evolved over centuries along with the enabling operational concepts, naval strategies, nation-states and attendant diplomacy. The system was thus able to gradually incorporate advances, slowly accommodate stresses, and, to some degree, resolve conflicts in a deliberate manner over time.

A key aspect of space is that the speed of advances in access and spaceborne capabilities has significantly outpaced the creation of guiding national-let alone international strategies and policies. The technological advances in space systems and increased reliance on them have created a space-enabled "critical infrastructure" that has not been matched by coherent supporting protection and loss-mitigation strategies, clearly articulated and accepted policies, and robust defensive capabilities. These gaps have created newfound concern domestically, confusion on the part of allies, and opportunities for misalignment and misperceptions on the part of potential adversaries. The need to rapidly, precisely, and effectively address all of these factors has created an environment of urgency to find mitigation strategies, fill policy gaps, and fund new capabilities. Done poorly, rapid efforts and expansive rhetoric can exacerbate existing tensions, pursue capabilities that add only marginally to system security, and increase the probability of misunderstanding or miscalculation on the part of potential adversaries. Well coordinated and properly executed, these efforts can meet real needs, add essential system security, and promote stability. These efforts must succeed. National security and global stability in space and on Earth demand it.

Space systems—systems with one or more components resident on Earth-orbiting satellites—are integral parts of the national and global information infrastructure. Some of these systems are essential parts of that infrastructure in that their functions either cannot be performed solely by terrestrial systems or can only be performed poorly and/or with great difficulty and expense by land, sea, or air-based substitutes. In the abstract, were all of the space systems suddenly to shut down, the global information infrastructure would cease to function as the world has come to expect; were the use of space to be denied in perpetuity, current information capabilities would be nearly impossible to reconstruct. Today, companies that operate space systems and the companies that use the services provided by those space systems accept the risk that they can be disrupted by both natural and man-made causes, and plan accordingly.[1] However, that risk is generally small. Were that risk to be perceived as being much larger, the business calculations would inevitably change, with potentially large consequences for both global commerce and daily life.

The list of human activities that are dependent on space systems contains most of the major functions that are vital to modern society, including trade and commerce; banking and financial transactions (from operations of major financial markets to minor retail purchases); personal, corporate, and government communications; agriculture and food production and distribution; power and water systems; transportation; news gathering and distribution; weather assessment and prediction; health care and entertainment. Were the world to suddenly be "without space," these would all seriously degrade or shut down entirely.

National security, in all its dimensions, is even more reliant on space systems. The U.S. military and other national security institutions are dependent on the reliable functioning of space systems in peacetime, crisis, and conflict, as are potential adversaries. U.S. national security depends on some of the same space systems that also serve important civil functions. There are, however, space systems that are solely for government use. Some of these are unique; some are vital; some are both vital and unique. National security functions provided by space systems range from the essential to the convenient, but the majority trend toward the former. The loss, or threat of loss, of secure communications; precise positioning, navigation and timing; and timely intelligence and surveillance of nearly every type, including missile warning, would dramatically and deleteriously affect the ability of the United States to conduct combat and other national security operations. Space systems enable everything from command and control to targeting and delivery of offensive capabilities to logistics and humanitarian relief. Space-based functions are also vital to most other nations and the global community in general, for maintaining stability during crises and more effectively addressing their own societal and security needs. Loss or degradation of those functions would increase the risk that a crisis would escalate into an unnecessary or unintended conflict.

Space systems are vulnerable to disruption from natural causes, from human activities that are not intended to damage or interfere with space systems, and from intentional attack. In the view of many, space has been, until recently, a "sanctuary" from intentional attack, but that sanctuary status has now eroded or vanished. Several nations have publicly demonstrated the ability to attack satellites on-orbit.

[1] Natural hazards to satellites include space weather, meteors, space debris, and cosmic rays. For additional information, see National Research Council, *Severe Space Weather Events: Understanding Societal and Economic Impacts: A Workshop Report*, Washington, D.C.: The National Academies Press, 2008.

However, any entity—government agency or private organization—that has the ability to launch a satellite to a precise orbital location, has at least latent capacity to attack a satellite by launching an object to an already-occupied location. Moreover, space systems are vulnerable to cyberattack by national or sub-national groups, including organized crime. The ground segments of space systems are themselves, or can be, vulnerable to hostile disruption or attack.

The importance of space systems to the United States and its allies and potential adversaries raises major policy issues. The demonstrated development of means to attack space systems by other nations—and the obvious potential for still more nations and perhaps non-state actors to develop such means in the future—raise practical problems that demand solutions. Moreover, there is an urgent need to address the increasing threat to vital U.S. space systems, a need that cannot wait until broader policy considerations have been fully developed. In the abstract, the United States would like to (1) have the means to deny anyone the ability to use space systems to support hostile actions against the United States; (2) maintain the ability to use space assets for national security purposes in peacetime, crisis, and conflict; and (3) ideally or idealistically, be assured that space remains a benign operating environment for all civil and commercial activities. These are not always mutually compatible.

Given the country's broad dependence on space for both civil and military activities, U.S. interests would appear to be served by a future in which there exist no means to unilaterally attack U.S. space systems without attribution and effective counters, or a future where space systems offer sufficient resiliency that such unilateral attacks are not effective in negating a space capability. However, given the dependence of potential adversaries on space systems in time of conflict, the interests of the United States may also be served by having the means to disable adversary space systems in time of crisis or conflict. Moreover, a number of means to attack space systems have been demonstrated or are postulated, and failure to protect against the use of such systems would put the United States at a significant disadvantage. While the United States may decide what space future it prefers, the United States is not the sole determiner of that future. U.S. actions will be constrained by what our potential adversaries—and even our friends—decide to do. Furthermore, frenetic innovation in the commercial space sector has the potential to be the main driver of change in the space domain. Put somewhat differently, the United States faces a short-term problem that needs to be addressed with urgency and it also faces a more complex, long-term problem.

In the short term, what should the United States do to counter the emerging, multi-faceted threat to U.S. national security space assets? Potential measures include hardening systems against known and predicted means of attack; establishing capabilities to mitigate the effects of successful attacks on U.S. space systems; expanding systems to detect attacks in progress, including confidently distinguishing attacks from other sources of failures; and reacting to them, implementing political-military means designed to deter attacks, and developing and deploying retaliatory systems and other means to hold adversaries' assets at risk. This is not just a matter of developing hardware; organizations, policies, doctrine, and operational concepts need to be modified or created in parallel. Policy issues include declaratory policies with regard to attacks on the national security space architecture, including commercial space systems that provide national security functions, as well as appropriate responses to attacks on significant commercial systems. Addressing this problem requires a clear understanding of the threat and the diverging time lines associated both with threat evolution and timely deployment of solutions.

In the long-term, failure to build means to protect U.S. space systems from attacks by existing weapons could significantly increase risks to U.S national security. However, relying only on developing defenses against known threats would cede the initiative to potential adversaries and would risk deploying tomorrow the counters to threats that were developed yesterday. Failure to at least consider responding to evolving threats by building U.S. systems to attack adversary space systems could also increase the risk to the United States and would certainly constrain U.S. options in response to an attack. However, simply following the path of "offensive defense" as a deterrence strategy also goes a long way toward fundamentally defining the future security situation in space. That situation, an unfolding arms race in space, may or may not be the future that best serves the long-term interests of the United States. Focusing only on the short-term problem in this manner is a reactive approach that, to some degree, cedes initiative to potential adversaries. While this may turn out to be the only pragmatic approach, it would also be in the interests of the United States to take a longer-term strategic view and assess what the United States wants the future of space to be, and what is our ability, in a global context, to help shape that future. So the priorities are as follows:

1. Develop a clear vision—or perhaps alternative options—of what the United States wants the future in space to be.
2. Understand the extent to which the United States can shape that future, and the extent to which the future is subject to actions and activities beyond the control of the U.S. government and its allies.
3. Identify and develop prudent methods to counter existing, evolving, and emerging threats to U.S. interests in space.
4. Assess those methods in terms of how they affect the future in space and the ability and the commitment of the United States to shaping that future.

CURRENT AND FUTURE SPACE-ENABLED CAPABILITIES

Space is congested, contested, and competitive, but space also touches most of the world's inhabitants. Directly or indirectly, everyone on Earth is affected and involved. Voluntarily or not, we are all now a part of the "space community." Space, along with its adjacency, cyberspace, has joined the domains of land, sea, and air as part of an interwoven, global, critical infrastructure providing essential information and connectivity. In the face of new risks and potential threats from many directions, we face uncertainties, over intentions and ambiguity as to outcomes that have slowed our efforts to "organize, train, and equip" to effectively position ourselves in response to either today's realities or tomorrow's possibilities. At its heart, this is a governance challenge. What is required is leadership—sustained, consistent, and effective.

There are a couple of important overlooked historical aspects of the evolution of national security space in the United States and globally that contribute to the current state of affairs and that are relevant to potential mitigating strategies: (1) Space from its earliest days was viewed as a sanctuary with little need for physical security. The distances to orbit, limited number of players, and the norms established during the Cold War all resulted in an institutional set of views and policies that have not been normalized to the current environment. (2) No full military conflict has yet been fought in, though, or from space wherein both sides have parity in space capabilities and dependencies. Thus, part of the apparent paralysis in the development of coherent space policy and doctrine with respect to space comes only from hypothesized scenarios, not from experience in battle. (3) Space has heretofore been largely a supporting "utility" to the warfighter under largely uncontested circumstances. This has prevented the evolution of a coherent, integrated operational art involving all warfighting domains with the space domain. (4) And the lack of global experience involving military use and negation of space capabilities has resulted in a lack of direct experience with how to value the risks and consequences that are central to deterrence.

PRESERVING NATIONAL SECURITY SPACE–ENABLED CAPABILITIES

The contested character of space need not and should not lead to conflict. With our newfound appreciation of the importance of space systems, we had better understand the significant threat to modern society that their loss represents and, in considering how best to respond, we appreciate both the urgency of the need and the depth of the challenge. While deterrence, in all its dimensions, must be part of our national strategy, a successful outcome nationally and globally requires all elements of diplomatic, intelligence, military, and economic domains to achieve outcomes desired nationally and acceptable globally. The sensitivity of space security discussions can complicate this cross-domain collaboration, but lessons can and should be drawn from successful de-confliction, if not cooperation, in other areas. Finally, the fact that the United States is unlikely to be fighting alone against peer or near-peer adversaries is important when considering appropriate space security strategies. The United States is inextricably linked and dependent upon its allies to fight with it—something that is well recognized by the establishment of interoperability standards (e.g., North Atlantic Treaty Organization standardization agreements) and common field training venues. Extending this paradigm to the space domain is critical for our overall net resilience.

FINAL THOUGHTS

It is important to note that the committee was not briefed on all of the details of classified programs, extant or planned, or on the allocation of recent funds identified in their support. Given the scope of the challenges, it is unlikely those resources currently budgeted will be sufficient or that all risks can be identified and eliminated within the 5-year program horizon. While some progress has been made in the development of common ground architectures and battle management command, control, and computers, much remains to be done. The collection of intelligence on emerging threats and capabilities must be timely and better identify sometimes disparate patterns of science and technology. On the operational side, expansion of the ability to quickly detect and corroborate the deployment or pre-employment of adversary systems, be they terrestrial or space-based, will be essential. In addition, clarifying operational authorities for national security space assets during a potential conflict extending to space will be needed. Some defensive concepts, such as disaggregation or distribution, must be rigorously evaluated to identify required cost (and the cost penalty imposed on adversaries), capability and resiliency compromises. From an intellectual and workforce perspective, the challenges do not appear to be insurmountable. The high-tech labor market will likely respond to a real and sustained program of development and deployment of national security space assets and continue to produce sufficient numbers of dedicated workers equal to the demanding tasks.

1

Context for Study

INTRODUCTION

It is not yet 60 years since the first artificial satellite was placed into Earth orbit. In just over a half century, mankind has gone from no presence in outer space to a condition of high dependence on orbiting satellites. These sensors, receivers, transmitters, and other such devices, as well as the satellites that carry them, are components of complex space systems that include terrestrial elements, electronic links between and among components, organizations to provide the management, care and feeding, and launch systems that put satellites into orbit. In many instances, these space systems connect with and otherwise interact with terrestrial systems; for example, a very long list of Earth-based systems cannot function properly without information from the Global Positioning System (GPS).

Space systems are fundamental to the information business, and the modern world is an information-driven one. In addition to navigation (and associated timing), space systems provide communications and imagery and other Earth-sensing functions. Among these systems are many that support military, intelligence, and other national security functions of the United States and many other nations. Some of these are unique government, national security systems; however, functions to support national security are also provided by commercial and civil-government space systems. Moreover, over the past quarter century the definition of "national security" has become expanded well beyond security against military attack to include protecting the security of the important functions on which the functioning of a complex modern society depends. For example, to the extent that the functioning of the national electric power grid depends on services provided by space systems, hostile actions against those services constitute a threat to national security.

In 1955, human life was not dependent on space systems. In 2016, most people's lives are touched every day in important and often fundamental ways by space systems. The loss of space systems in general, and any of many individual space systems specifically, would be highly disruptive. Both the extent of this shift and the rapidity with which it has occurred are noteworthy. The projection of this trend into the future is a matter of considerable uncertainty and debate. Some project continued rapid expansion; to others, the growth has been asymptotic and will now slacken. Yet others opine that the trend will reverse as the risks of being overly dependent on space become clear.

A major factor behind these risks is the vulnerability of space systems to disruption. Like all human activities, there are risks from natural disasters and the detritus of human activity—primarily accumulated orbiting space debris. However, the greatest vulnerability to U.S. space systems is thought to be an intentional hostile action by another actor. These changes as they apply to society in general have been mirrored in the U.S. military. Circa 1980, only a few functions relied on space systems; specifically, strategic nuclear command and control, weather and climate monitoring, and military and intelligence communications. Today, nearly all activities at all levels in civil society as well as defense depend in some way on space functions, both in peacetime and during conflict.

Defense and intelligence systems have been in space since the earliest space systems. The notion of denying adversary space assets in time of crisis was not far behind. Official documents of the Ford Administration (1974-1977) contain assertions that Soviet space systems could serve strategic, operational, and tactical purposes during a conflict, as well as observations that the United States should be prepared to deny the Soviet Union those systems as necessary.[1] Similar attitudes toward U.S. space systems were attributed to the Soviet Union. During the 1960s and 1970s, both the United States and the Soviet Union were developing means to attack satellites. In 1959, the United States had attempted to intercept the Explorer 5 satellite, but failure of test equipment precluded assessing the degree of success. An adapted version of the nuclear-armed Nike Zeus missile was deployed on Kwajalein atoll from 1962 to 1966, when it was replaced with a variant of the Air Force Thor program that remained operational until March 1975. The Soviet Union began experimenting with antisatellite (ASAT) systems in approximately 1960 and conducted successful tests of a co-orbital interceptor in 1967 and 1968. A system based on these tests was declared operational in early 1973.

[1] U.S. Department of State, Office of the Historian, *Foreign Relations of the United States: Volume E-3, Documents on Global Issues, 1973-1976*, released December 18, 2009.

By the late 1970s, the situation was considered sufficiently serious for the two nations to engage in exploratory discussions about the possibility of negotiated controls of ASAT weapons. Even in 1978, this was a difficult undertaking, despite the fact that only two nations had ASAT capabilities, and between them accounted for the lion's share of all space systems, almost all of which were government owned. Recognized means of attack at the time included (1) mechanical/kinetic direct ascent interceptors; (2) co-orbital "space mines" (as they were then termed); (3) nuclear warheads in space; (4) directed energy weapons fired from the ground, aircraft, manned spacecraft, and other satellites; (5) jamming; and (6) malicious signals inserted into housekeeping and other communications to satellites (now called cyberattack). Even with only two ASAT-capable nations, attributing damage to a satellite was viewed as uncertain, as was, in the case of some forms of attack, distinguishing attack from innocent malfunction or the result of natural incidents. Emerging multinational and commercial involvement in space systems was a complicating factor.

Today, an increasing number of countries have the ability to build and launch a vehicle capable of reaching orbit, including the United States, Russia, China, India, Iran, South Korea, North Korea, and the member states of the European Space Agency. However, governments and their large budgets are no longer the only drivers of space activities, whether civil, commercial, or even security-related. Rather, a new set of non-state space actors have become foundational and catalytic elements of space activity. While the United States and Russia have maintained their lead in space, other countries are now also able to leverage the strategic and tactical advantages provided by space capabilities. Just 50 years after the launch of the first commercial satellite, Intelsat 1 (1965), the dynamics of space activity continues to shift towards the commercial use of space for consumers and businesses as well as traditional government customers.

SPACE-ENABLED CAPABILITIES ARE INCREASINGLY SHARED

It is both convenient and informative to have the "big picture" on the fundamental science and engineering of the "space systems" that are the main focus of this report. These space systems consist of satellites that orbit Earth and their associated ground-based systems (e.g., a DirecTV receiver located in a home). Earth orbiting satellites are information nodes in a larger global network. From this perspective, the vast majority of operational satellites in orbit are common in their basic design and operational principles, in that satellites:

1. *Collect information.* Intelligence, surveillance, and reconnaissance (ISR), missile warning, and weather and environmental satellites collect light at a variety of wavelengths from visible to radio frequencies. Communications satellites

(COMSATs) likewise collect radio transmissions. Such transmissions can be sent from the ground, air, sea, or, in the case of the Iridium system, from another satellite. Positioning, navigation and timing (PNT) satellites (e.g., GPS) receive from the ground a vital piece of information—their precise position in space.

2. *Process and/or store information.* The information collected on board a spacecraft is converted and stored in the satellite. ISR, weather, and environmental satellites convert light into digital images. Conventional COMSATs convert radio signals received at one frequency to a different frequency in preparation for retransmission. Newer and more sophisticated COMSATs may actually handle incoming information much like an Internet router, ensuring packets of digital data are given the appropriate path in a network to get to the intended user(s). Navigation and timing satellites package the satellites' own precise position along with precise time provided by an onboard atomic clock. Because of the inability of many satellites to transmit everything collected in real time, much of the information collected must be stored onboard for subsequent transmission.

3. *Distribute information.* The information that has been collected, processed, and/or stored is then retransmitted to users on the ground, air, and/or sea below. In cases such as satellite radio, satellite TV, and GPS, the information is transmitted directly to an end user that has receiving equipment. In other cases, such as weather satellites and personal communications systems (e.g., Globalstar), the information is first transmitted to a ground station that then routes the information through a variety of networks, which could include other COMSATs, to eventually get to the end user.

This basic view of the theory of space system operations offers more clues into the primary utility of satellites. The first thing satellites do is collect information. Space provides a unique position for such information collection. First, much more area of Earth can be seen from space at any one time. Hence, one COMSAT can see as many ground users as could the equivalent of millions of land-based cell towers. Second, space provides the ability to collect information from areas in which access is otherwise denied, physically and/or politically.

This framework also provides a basic primer on how threats can deny, disrupt, and/or degrade information flows to, through, and from space systems: interfering or destroying the ability to either collect, process/store, and/or distribute information can stop the flow of information to end users. Specifically, threats can do the following:

1. *Deny, disrupt, and/or degrade information collection.* Radio signals to COMSATs from the ground can be overridden by jamming systems. Jamming systems are basically radio transmitters tuned to the same frequency as the signals being sent to a satellite above. If the received signals from the jamming system are more

powerful than the actual signal, the actual signal cannot be received properly by the satellite. Electro-optical ISR systems can likewise be "blinded" by lasers directed at them.

2. *Deny, disrupt, and/or degrade information processing and/or storage.* A main threat here is a physical attack on the satellite itself. An ASAT weapon can be designed to physically deny, disrupt, degrade, and destroy a satellite. By taking out the satellite, the main link in the chain of information is broken.

3. *Deny, disrupt, and/or degrade the distribution of information.* As most information is distributed via radio waves, this would involve a radio jamming approach somewhat like that used for interfering with the collection of information. However, this approach is typically done from the air or ground and, as such, is limited in geographic scope. An example of this type of attack is GPS jamming, in which small ground-based transmitters can prevent GPS users from receiving GPS satellites signals in localized areas.

This information-centric framework has another important implication directly related to cyberspace. Cyberspace is, to put it plainly, the domain of worldwide information flows between humans and machines that is enabled by a complex system of computing, switching, storage, and relay devices and infrastructure (e.g., fiberoptic cable). In this view, the space systems are inherently a component of, not separate from, cyberspace.[2] Satellites are nodes in a network, and their value is derived from their ability to collect and disseminate information on the network. This is not mere semantics: As part of cyberspace, space systems can be equally threatened from cyberattack. For instance, a virus can interrupt the function of a satellite handset. Likewise, a virus placed into a satellite could prevent the proper onboard processing of information or the proper operation of the satellite itself.

Today, customers of space goods and services, whether civilian or military, seek access to new and innovative technologies that make life, work, combat, and governance more connected, accessible, efficient, and transparent for them. Both the DirectTV user and ISR user are demanding more data, bandwidth, and digital accessibility to fulfill their needs. This has resulted in increased demand for all types of bandwidth, in particular, mobile services that included space-based assets. Government-owned capabilities and budgets alone have not been able to, and cannot, meet the growing consumer appetite, thus creating opportunities for commercial actors to help governments meet civilian and military demand. While worldwide government spending had shrunk to 24 percent of total annual space revenues in 2014, compared to the 35 percent it held in 2006, the commercial side

[2] The International Space Station (ISS) is an example of a space system that, in addition to being a part of cyberspace, serves as a laboratory. In the future, space servicing robots and logistics supply vehicles moving satellite fuel from one place to another will also transcend this cyber-centric view.

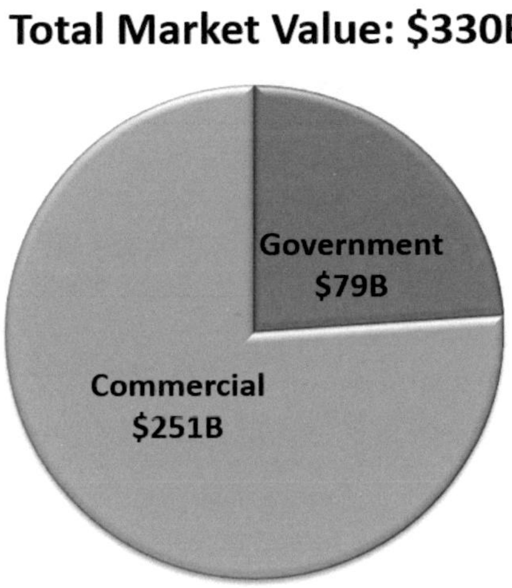

FIGURE 1-1 Global space economy, 2014. SOURCE: Data from Space Foundation, *The Space Report: The Authoritative Guide to Global Space Activity*, Colorado Springs, Colo., 2015.

of space has experienced nearly $100 billion in growth, almost doubling the amount of money entering space for private use.[3] Two primary segments of the commercial space market, communications and remote sensing, are expected to grow to $35 billion and $6 billion, respectively,[4] to cater to the demand from nearly four billion mobile device users and over three billion Internet users.[5] All users are in turn reliant on the position, navigation, and timing capabilities provided by government-operated space-based satellite navigation systems. Simply put, demand for commercial satellite services is not just growing, it is accelerating (see Figure 1-1).

---

[3] Space Foundation, *The Space Report: The Authoritative Guide to Global Space Activity*, Colorado Springs, Colo., 2015, p. 39.

[4] Multiple sources: International Telecommunication Union, Measuring the Information Society, Place des Nations, Geneva, Switzerland., 2013; Cisco Forecast White Paper, https://www.cisco.com/c/dam/en_us/about/ac79/docs/innov/IoE_Economy.pdf; World Bank Global Indicators 2013; World Bank Group, ICT for Greater Development Impact World Bank Group Strategy for Information and Communication Technology 2012-2015, June 15, 2012; Machina Research: Future of M2M market, http://www.telecomengine.com/sites/default/files/temp/CEBIT_M2M_WhitePaper_2012_01_11.pdf; AT Kearney GSMA 2013 global report, https://www.atkearney.com/documents/10192/760890/The_Mobile_Economy_2013.pdf; International Data Corporation Worldwide Quarterly Mobile Phone Tracker, https://www.idc.com/tracker/showproductinfo.jsp?prod_id=37; Geospatial World conference January 2014, http://geospatialworldforum.org/2014/; EuroLinker 2014 commercial imagery report; Cisco, Information Technology Update 2015, http://www3.weforum.org/docs/WEF_Global_IT_Report_2015.pdf.

[5] We Are Social, Ltd., "Digital, Social, and Mobile Worldwide in 2015," 2015, http://wearesocial.com/uk/special-reports/digital-social-mobile-worldwide-2015.

The increasing demand for commercial space services, with its primary drivers being the communications and remote sensing fields, illustrates the growing importance of space-based technologies for civilian use. The current information revolution is creating new opportunities to utilize this expanded ability, fueling a growing demand for readily available space-based technologies. As the launch of new satellites expands capabilities, new uses are being imagined and explored for the next generation of launches (see Figure 1-2).

Smartphones, tablets, and the "cloud" represent the current generation of products of the information revolution. These capabilities are the facilitators of globalization and will continue to create a world that is increasingly interconnected. An architect in Bangalore can send a picture of a new data center facility to colleagues sitting in Silicon Valley with the touch of a button from an iPhone. Through satellite imagery, a geospatial intelligence analyst can pinpoint vulnerabilities in the Philippines' Special Economic Zones, where oil smuggling is rampant. By deploying Broadband Global Area Network (BGAN) satellite terminals, the Brazilian Electoral Commission can easily access voters in rural and remote locations with unprecedented speed and accuracy.[6] Globalization and the accompanying information revolution enable a level of interconnectivity and convenience unimaginable to even futurists during the first generation of satellites.

THE ACCESSIBILITY OF SPACE

The global population is predicted to grow to approximately 8.3 billion by the year 2030, while people will consume, access, create, and share digital information between one another, and with businesses, infrastructures, and machines at an increasing rate. It is expected that the world will become increasingly hyperconnected—a state of amplified interconnectivity between people, business organizations, and governments, which transcends the physical limitations of geographical boundaries. Rules will change and power will shift in industries and markets, challenging the current way wealth is created and distributed globally. The consumer's purchasing power can be expected to be amplified and consumers will use this power to press for even larger technological growth to meet demands in their daily life, as well as business, military, and governance activities. More subtly, consumers will expect that they can trust these information-driven goods and services to be always available and reliable, similar to their expectations in regard to electricity and water utilities that are relied upon by billions of people today. The information collected and shared, as well as the enabling devices and technologies, is often dependent on commercially provided space-based products

[6] MarketWatch, "Smartmatic to Provide Election Services in Brazil," press release, June 11, 2014, http://www.marketwatch.com/story/smartmatic-to-provide-election-services-in-brazil-2014-06-11.

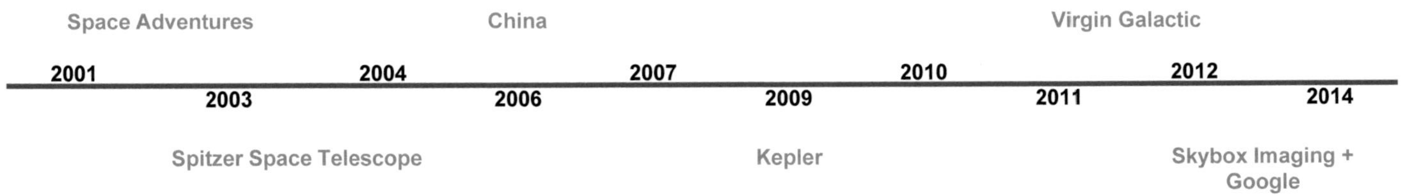

Date Details

2001 – American Dennis Tito became the first tourist in space after paying the Russian space program $20 million
2003 – NASA launched the largest-diameter infrared telescope in space, the Spitzer Space Telescope
2004 – President Bush proposed a new program that would send humans back to the moon by 2015 and establish a base to Mars and beyond
2006 – NASA spacecraft, Stardust, returned safely to earth in with the first dust ever collected from a comet
2007– China downs one of its weather satellites with a ground launch missile
2009 – NASA launces Kepler spacecraft to search for planets outside our solar system
2010 – Virgin Galactic announced the successful first manned glide flight of the VSS Enterprise
2011 – Last American space shuttle, Atlantis, launched to restock ISS supplies and equipment
2012 – SpaceX launched its Dragon C2+ mission to resupply the ISS
2014 – Google buys satellite imagery company Skybox Imaging

FIGURE 1-2 Big moments in space history: 2001-2014.

and services, in turn driving the commercial space industry's evolution. While new technologies signal to companies where development in space should occur, often times the infrastructure put in space is utilized in unforeseen ways in response to newly developed demands.

Space systems have in many ways become a vertical extension of terrestrial networks, or, looked at in another way, high-altitude components of the increasingly integrated global information network. Space-based sensors can collect information available only from the vantage point of space, and communications satellites are an efficient means to effect global distribution of data and information. It has been estimated that in 2014, space-enabled systems created nearly $200 billion in global direct economic activity, which was up 300 percent from 2000.[7] However, these monetary figures do not fully measure the impact space has on the U.S. economy and security. Finally, even non-state actors, such as transnational criminal and violent extremist organizations, are making use of space-enabled capabilities, including the use of the Internet for recruitment, funding, and planning.

DOMESTIC AND INTERNATIONAL CONSUMER MARKETS

The majority of the purchasing power base for space has already shifted. Today, end-users drive more of the decisions and shape the design and production of capabilities and products across all industries and markets, including commercial space. The voracious consumer craving for bandwidth, access, and security drives the communications market and presents tremendous opportunities for additional commercial business, particularly in communications and remote sensing. Combined market demand is estimated to grow 20 percent from now to 2024, providing a value of over $40 billion.[8] The growth trajectory of these commercial markets and the industry as a whole is, and will be, directed by the consumer's behavior over the next half century and industry's response to that behavior (see Figures 1-3 and 1-4).

Increased demand for mobility, bandwidth, and interconnectivity exists not only in advanced economies but also in developing countries. Because of this increased demand, alternatives to terrestrial infrastructure solutions are being pursued. New satellite architectures to deliver communications and imagery are being developed to meet growing consumer demands. A burgeoning middle class and a growing population of independent consumers in the Asia-Pacific, Arabian Gulf,

---

[7] Doug Loverro, DASD for Space Policy, "Defending Space," briefing for USD (AT&L), May 6, 2014.

[8] Multiple sources: ITU Forecast Report 2013, Cisco forecast whitepaper World Bank Global indicators 2013, and World Bank trends in telecom 2012, Machina Research: Future of M2M market, AT Kearney GSMA 2013 global report, International Data Corporation Worldwide Quarterly Mobile Phone Tracker, Geospatial World conference, January 2014, EuroLinker 2014 commercial imagery report, and Toffler Associates research and interviews.

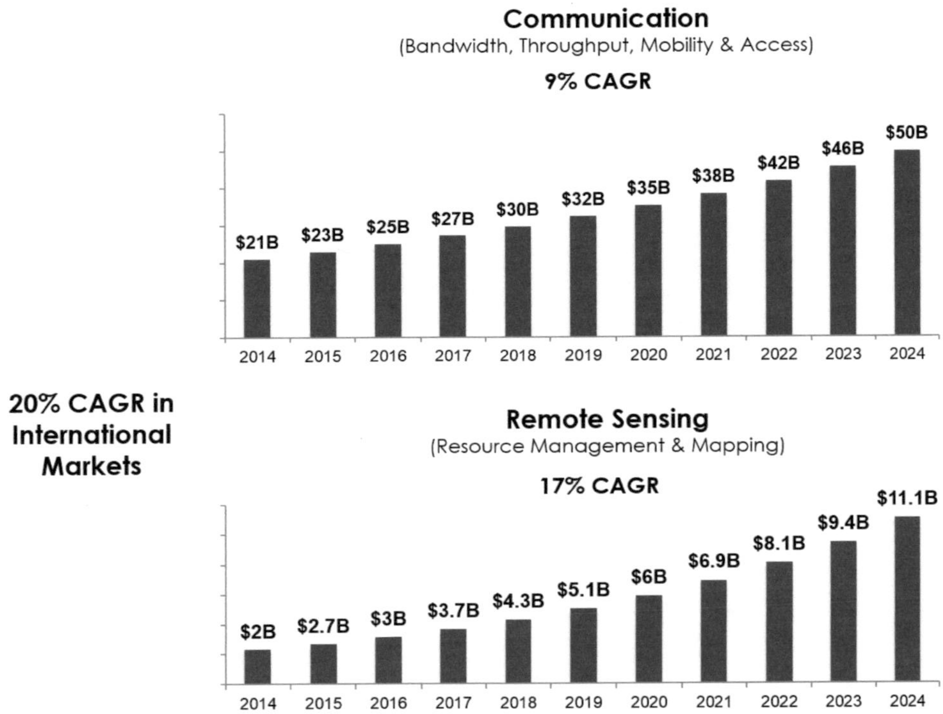

FIGURE 1-3 Growth of commercial space communications and remote sensing markets (2014-2024). NOTE: CAGR, compound annual growth rate. SOURCE: Data from Space Foundation, *The Space Report: The Authoritative Guide to Global Space Activity*, Colorado Springs, Colo., 2015.

and West and South African countries are untapped markets with a substantial amount of underserved people who have disposable income and who are ready to join in on the hyperconnectivity revolution. The needs of this new user base mimic those in developed countries with increased access to networks. Unfortunately, this new citizen-consumer demographic is disadvantaged owing to geographic limitations in regional infrastructure and politics. Satellite operators, space-based technology providers, and other commercial remote-sensing and communications competitors that move into these emerging international markets will encounter consumers who want immediate access to greater amounts of data requiring more bandwidth. These consumers are ready to spend on mobile devices and other space-based technologies, as the global devices market is projected to sell over four billion consumer mobile-to-mobile devices by the year 2030. However, the growth in demand for space-based technologies in international markets will converge in areas where there are existent geopolitical tensions. For example, within the Asia-

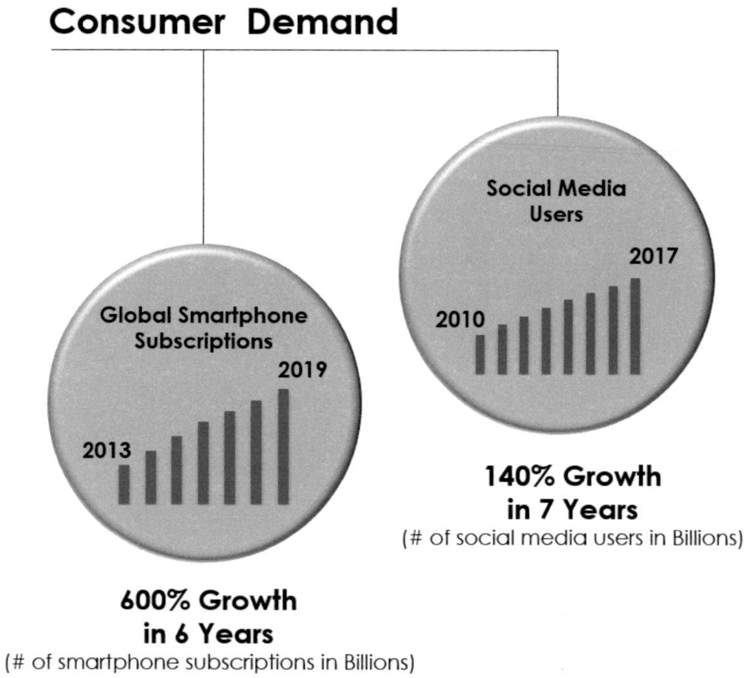

FIGURE 1-4 Consumer demands to stay connected through devices is driving new space capabilities. SOURCE: Data from Cisco, Information Technology Update 2015.

Pacific region, there were about 3.6 billion mobile device subscriptions in 2014—51 percent of the world's total.[9] As U.S. and international commercial companies expand into those markets, they will need to partner with local companies, possibly state-owned, and local governments to provide services to the local consumers. As these commercial companies provide services vis-a-vis regional satellites, the likelihood of risk to U.S. military operations may increase.

The accelerating spread of connectivity presents challenges. As powerful global forces emerge, smaller countries and non-state actors are now able to access advanced technologies and partake in the information revolution. Therefore, technology can also be employed by illicit non-state actors, or groups seeking to advance their agendas, with the intention of threatening a country, region, or in the case of the United States, the space capabilities on which our security and economy

[9] Asia Trend Bulletin, "10 Asian Trends for 2015," December 2014-February 2015, http://trendwatching.com; ITU, 2014.

depend. China and Russia are developing counter-space capabilities that can be used to disrupt U.S. capabilities and by doing so weaken U.S. global stature. To face these growing threats, the United States will need to develop new approaches with the appropriate technological tools to maintain space-based capabilities and technologies. Lethal uses of space-based technologies are only a fraction of the overall consumption and demand for such technologies. The use of space-based technologies cannot be eliminated if for no other reason than that consumers need these technologies for their daily personal and professional livelihoods.

GOVERNMENT AND COMMERCIAL SECTORS

As the citizen-consumer attracts a majority of the commercial space operators' focus, the U.S. government's prominent role in driving the space market continues to diminish owing, in part, to ongoing budget constraints. Looking back to the partial federal government shutdown and sequestration in fiscal year (FY) 2013, there was a nearly 10 percent decrease in federal spending on space from 2012, with only a slight increase (3 percent) in 2014.[10,11] The trend of downward budget pressures makes it difficult for the U.S. government's space agencies to keep up with the rising space economy's demand and satisfy our nation's needs.

Some functions are shared between civil and military users; domestic and foreign users; government and industry. For example, weather forecasting, especially that related to potential catastrophic natural events, remains deeply dependent on both low Earth orbit (LEO) and geostationary orbit (GEO) meteorological satellite systems. Similarly, new commercial imagery systems, such as PlanetLabs and Terra Bella 2, hold the promise of providing temporally and spatially persistent global geospatial awareness capabilities previously unavailable from even the most advanced U.S. government space systems. Likewise, future Internet constellations, such as OneWeb, have the potential to ensure that the remaining unconnected populations of Africa and Asia become a part of the global "informationized" community. Commercial and military communications spacecraft in the GEO support global communications connection to users on land, in the air, and at sea (Figure 1-5).

The government's budget problem is shifting not only its space activity, but also the military consumer's choices of technology. Given the access, efficiency, and enhanced capabilities that systems such as Remotely Piloted Aircraft (RPA) or Unmanned Aerial Systems (UAS) provide, the U.S. government is transitioning to being a consumer of readily available, commercial space-based capability to support

[10] Space Foundation, *The Space Report: The Authoritative Guide to Global Space Activity*, Colorado Springs, Colo., 2014, p. 56.

[11] Space Foundation, *The Space Report*, 2015, p. 39.

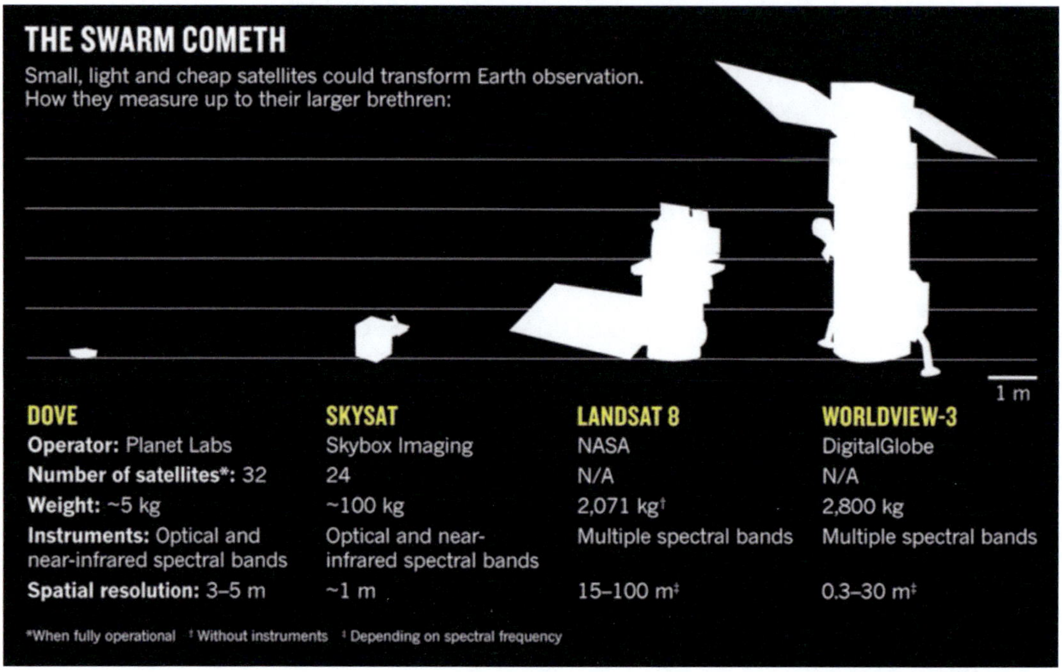

FIGURE 1-5 New commercial remote sensing satellites are small. SOURCE: Reprinted by permission from Macmillan Publishers Ltd: *Nature*, D. Butler, Many eyes on Earth, *Nature* 505:143-144, 2014, copyright 2014.

military as well as many other applications. This is particularly true for Department of Defense (DoD) capabilities that demand an advanced communication platform for missions abroad. Warfighters expect to access data anytime, anywhere from a mobile device just as they do in civilian life. During ISR collection missions in remote areas, military personnel and vehicles will need more bandwidth to operate a new generation of increasingly more prevalent tools, such as UASs, which utilize high bandwidth levels. As such, the federal government's demand for bandwidth is projected to grow from $3.5 to $5 billion in the next 15 years.[12] Such shift in behavior of the military consumer as well as the civilian consumer illustrates the convergence of the once distinct sectors of space (see Figure 1-6).

[12] Toffler Associates analysis and interviews. DoD Budget Data Deltek.

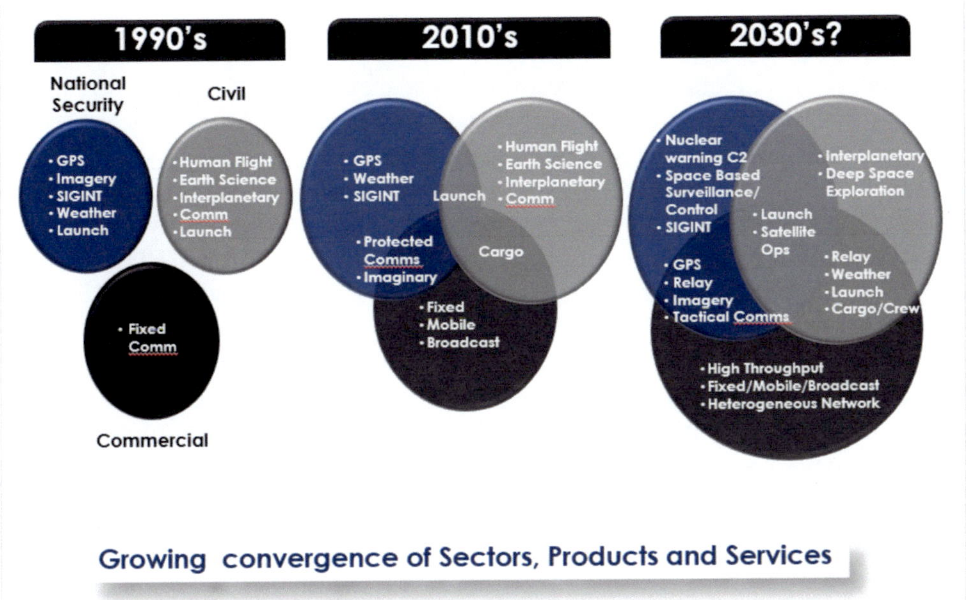

FIGURE 1-6 Evolution of space sectors.

DoD currently depends on commercial satellite communications systems for about 40 percent of its communications needs.[13] Coming from the other direction, the government's GPS, which started as a military system, has become an inherent and enabling component for civilian transportation, communications, and remote sensing networks.[14] The precision navigation capabilities from this space service will become even more vital as new air-space management systems (i.e., Automatic Dependent Surveillance-Broadcast) and nascent autonomous automobiles place critical reliance on GPS service signals. Use of GPS permeates a broad array of civil activities—especially commercial activities—ranging from point-of-sale financial transactions, through air, sea, and land transport, to cell phones, personal navigation, and various forms of recreation.

[13] Department of Defense, Chief Information Officer, "Satellite Communications Strategy Report," in response to Senate Report 113-34 to accompany S.1197, National Defense Authorization Act for Fiscal Year 2014, August 14, 2014.

[14] Warfighters expect to access data anytime, anywhere from a mobile device just as they do in civilian life. This expression might not be possible in a conflict against a peer adversary.

CONSUMER DEMANDS HELP DRIVE INNOVATION IN SPACE

Existing developers and new commercial space market entrants are exploring creative, niche technologies that enable their businesses to operate in previously unforeseen ways. These new technologies are enabling mission operations that use cost-effective value-added models. For instance, in 2013, in collaboration with British Telecom and Ericsson, Intelsat demonstrated the first-ever live sporting event recorded in 4K format, a type of ultra-high definition television (U-HDTV) that delivers footage with four times higher resolution than high-definition television (HDTV).

A new trend is the increasing use of smaller satellites that, because of their lower costs, remove barriers to entry for smaller companies (and nations) or create new market opportunities for larger corporations. A typical geostationary satellite might have a mass of 2,000 kg or more. "Smallsats," on the other hand, have a mass of between 100 and 500 kg. Microsatellites have masses of 10 to 100 kg, while nanosatellites have masses between 1 and 10 kg. The latter class of satellite includes the "CubeSat" class. A CubeSat is a cube 10 cm on each side, originally created to enable universities to build, launch, and test satellites using course budgets and timelines. Today, commercial companies, such as PlanetLabs, have turned to CubeSats to propel their business plan to image every portion of Earth once every 24 hours. CubeSats offer the advantage of low cost (tens to hundreds of thousands of dollars) and short production schedules (on the order of months and even weeks).Their disadvantage is of course their smaller size. Larger satellites are able to collect more information (e.g., light) at any one time. For an imagery satellite, this means a larger satellite can have better resolution for an image (PlanetLabs satellites have image resolutions of about 3 meters, as compared to submeter resolution for a larger commercial imagery satellite). To compensate for their small size, smaller satellites must be placed in lower orbits to be closer to the places they image or, in the case of COMSATs, to the origin of the transmitter(s). Getting closer reduces the area of Earth a satellite can view, so more satellites are needed if the system needs total Earth coverage at one time. OneWeb, a company that has backing from several larger groups, currently plans to create a LEO constellation of over 600 microsatellites (a notable space traffic management concern), each with a mass of about 150 kg and a price of about $500,000. The launches of nano- and microsatellites (combined) are increasing: From 2000 to 2012 roughly 20 such satellites were launched annually. In 2013, this total increased to over 90. In 2014, 158 nano- and microsatellites were launched, of which 107 were operated by commercial organizations.[15] This will remain a highly dynamic business area

[15] E. Buchen, "Small Satellite Market Observations," 29th Annual AIAA/USU Conference on Small Satellites, Logan, Utah, August 2015.

and area of interest for national security. Continuing advances in microelectronics will enable further advances in small satellites. Business opportunities in a variety of data products and communications capabilities will drive demand. Launch access will continue to constrain supply (most micro- and nanosatellites "hitchhike" their way to orbit, sharing rockets with larger satellites, although OneWeb plans on dedicated launches).

With more than 1,000 active satellites now orbiting Earth, space satellite manufacturers and space network operators understand that commercial space competition today differs substantially from in the past.[16] Competition is now more diverse, agile, and responsive. A wide range of products and services is offered, which may be space informed but not always space dependent.

THE VITALITY OF SPACE

Today, most people's lives depend every day in important ways on space systems. The loss of space systems in general, and certain individual space systems specifically, would be highly disruptive. Both the extent of this dependency and the rapidity with which it has occurred are noteworthy. However, the projection of this trend into the future is a matter of considerable uncertainty and debate. Some project continued rapid expansion and increased dependency; to others the growth has been asymptotic and will now slacken. Yet others opine that the trend will reverse as the risks of being overly dependent on space become clear. A major factor behind these risks is the vulnerability of space systems to disruption—either intentional or unintentional. Like all human activities, there are risks from natural disasters and the detritus of human activity—primarily accumulated orbiting space debris. A growing vulnerability for space systems is from intentional hostile actions by another actor. These changes as they apply to society in general have been mirrored in the U.S. military. Circa 1980, only a few military functions—specifically, ballistic missile warning, weather monitoring, and certain military communications—were reliant on space systems. Today, nearly all activities at all levels depend in some way on space functions, both in peacetime and during conflict.

In many ways, the "market" for space might have very little to do with space itself. The space industry may be compelled to reassess how space is used in today's complex geopolitical environment. As the space industry evolves, it is possible that it will be less an "industry" in of itself and more a domain occupied and leveraged

---

[16] As an illustration, according to the Union of Concerned Scientists, as of January 2015 there were 1,265 operational satellites: 669 in LEO, 465 in GEO, 94 in MEO, and 37 in elliptical orbits. Of these, 528 are U.S. satellites, including 160 military satellites and 121 belonging to other government entities (Union of Concerned Scientists, "UCS Satellite Database," http://www.ucsusa.org/nuclear-weapons/space-weapons/satellite-database, released February 1, 2015).

by other industries to serve purposes of research, information transmission, and national security. For example, the drive for big data is creating new demands for geospatial data and the merging of GPS with remote sensing capabilities. Currently, much of the business related to space is about leveraging space infrastructure to provide Earth-based services.

NATIONAL SECURITY USES OF SPACE

There are over 1,000 operational satellites in Earth orbit providing a wide array of critical functions. In addition to functioning satellites, there exist a large number of additional man-made objects in orbit resulting from previous operations, satellite failures, inadvertent collisions, and ASAT demonstrations. It is estimated that more than 21,000 space objects exist that are larger than 10 cm, and over 500,000 objects exist that are between 1 and 10 cm. Tracking this large number of space objects complicates space situational awareness, and inadvertent collision with space debris is a very real concern, especially in LEO. The orbits of these satellites can be categorized as LEO, medium Earth orbit (MEO), GEO, and highly elliptical orbit (HEO).

Low Earth Orbits and Functions

LEO refers to satellites orbiting Earth at altitudes higher than approximately 150 km and less than 2,000 km. These orbits are typically circular with orbital speeds between 7.8 km/s and 6.9 km/s and periods between 85 min and 130 minutes. One widely used orbit for remote sensing of Earth's surface is the Sun-synchronous orbital, which is a polar orbit in the plane containing the north and south poles and the center of the Sun. This orbit has the advantage that Earth rotates "underneath" the orbit, resulting in global coverage. LEO satellites are used for Earth observation, communications, and orbiting manned spaceflight. For example, the ISS is in orbit at an altitude of approximately 400 km, the National Oceanic and Atmospheric Administration (NOAA) operates a series of weather satellites in polar orbits at 850 km altitude, and Iridium operates a series of 66 communication satellites in orbits at 750 km altitude. OneWeb has announced plans to place approximately 700 satellites in 20 orbital plans at altitudes of 800 km and 950 km to provide high-speed Internet access across the globe.

Medium Earth Orbits and Functions

Satellites operating in MEO are located between 2,000 and 35,000 km. Satellites used for PNT functionality are primarily operated in MEO. As an example, the U.S. GPS operates a constellation of approximately 32 satellites in six orbital planes at an altitude of approximately 20,180 km, which results in a period of one-half of a sidereal day.[17] The Russian GLONASS system operates a constellation of 29 satellites (24 in the nominal constellation) in three MEO planes at an altitude of 19,100 km. The Chinese BeiDou system is planned for 27 MEO satellites at an attitude of 21,150 km, in addition to five geostationary satellites. The European Galileo system plans for a nominal constellation of 30 MEO satellites at an altitude of 23,200 km. Thus, approximately 118 satellites operate at altitudes between 19,100 km and 23,200 km, providing PNT capabilities to a wide variety of military, civilian, and commercial users.

Geosynchronous Earth Orbits and Functions

GEO refers to circular orbits with a period of one day, which corresponds to an altitude of 35,786 km and a speed of 3.07 km/s. One particular orbit of interest is the geostationary orbit, which is a geosynchronous orbit in the equatorial plane. Satellites in these orbits have the advantage of remaining over one spot on Earth as both Earth and the satellite rotate about Earth's axis at the same rate. Because geostationary orbits appear to remain fixed over a given location on Earth, they are extremely useful for communications, television and radio broadcasting, and Earth observation. At present there are approximately 95 geostationary commercial satellites in the GEO belt, with some operating with as little as 1/10th of a degree angular separation, which translates to approximately 73 km minimum spacing between satellites. As the geostationary belt offers many advantages, this portion of space is considered prime real estate. Space systems developers are encouraged to provide for "disposal" of their satellites at their end of life by maneuvering the satellite to a disposal orbit located approximately 300 km above the geostationary orbit. Once in this orbit, the nominally "dead" satellites drift slowly westward relative to the geostationary belt due to their higher altitude and slower velocity.

Highly Elliptical Orbits and Functions

HEO satellites, which travel above the horizon at high latitudes for significant fractions of their' orbit, have been used for communications, signal collection and Earth monitoring. This type of orbit has the advantage of requiring less insertion energy than geosynchronous orbits, but it comes with the disadvantage that antennae must be steerable to maintain Earth pointing. These orbits were originally explored by Russia to provide high-latitude communications coverage over its landmass. These so-called Molniya orbits (named after their early Molniya

[17] A sidereal day is the time it takes for Earth to rotate once relative to distant stars. The mean sidereal day is 23 hr, 56 min, 4 s.

communication satellites, developed in the mid-1960s) were highly elliptical and inclined at 63.4 degrees with apogee at approximately 40,000 km and perigee at approximately 1,000 km, resulting in a 12-hour period with apogee occurring over approximately the same high-latitude point on Earth. A second class of HEO is the Tundra orbit, which has the same 63.4 degree inclination but a higher apogee, such that the period of the orbit is one sidereal day. The Sirius Satellite Radio system operates three HEO satellites to maintain two satellites over North America to provide its broadcast service.

FINAL THOUGHTS

This context-setting chapter is intended to illustrate the degree to which space is no longer a purely military or intelligence-gathering domain exclusive to major world powers. Indeed, the explosive growth of the global commercial space market, coupled with the increase in government and civilian presence in space, reflects ever increasing civilian, commercial, and government dependence on space. Chapter 2 describes a whole-of-government approach to space, emphasizing the use of all elements of national power with the goal of ultimately deterring future conflicts in space.

2

Selected Issues Related to National Security Space Defense and Protection

INTRODUCTION

As noted in Chapter 1, to a greater degree than many of us realize, populations around the world, both in developed and developing areas, depend on space assets for convenience, quality of life, and resilience in the face of disaster. Satellites have brought nearly instant communication to places and people previously isolated by geography or a lack of ground-based infrastructure. Space-enabled communication has been particularly beneficial in areas of the developing world where landlines have not been laid or are unreliable, as well as to increasingly mobile populations. Data collected from space-based platforms have also brought new levels of precision to weather modeling, enabling disaster warnings that save lives.[1] Space systems are used to aid in search and rescue efforts and as a source of emergency communications in disaster areas. They are critical features of the Global Positioning System (GPS) that assists navigation for commercial vessels and aircraft, and that provides the precision timing and positioning essential to managing large

[1] An illustrative success story is found in the Bangladesh *Flood Forecasting and Warning Centre*, a group of 22 professionals in Dhaka who use National Aeronautics and Space Administration (NASA) and National Oceanic and Atmospheric Administration (NOAA) satellite imagery, plus mobile phone reports, to track the progress of the severe annual flooding and monsoon rains that caused so many casualties in the past. With more warning time, people have been better prepared for the onslaughts in recent years. See the Flood Forecasting and Warning Center website, http://www.ffwc.gov.bd/index.php/about-us/, for real-time maps and evidence of what a relatively small group can do.

computer networks and global financial flows, from secure wireless stock trading to automatic teller machines.

Also, to a greater degree than many Americans realize, space systems are a critical component of the national security services we now take for granted: Communication satellites, space-based imagery, positioning, navigation, and timing (PNT), and signals intelligence provide navigation and mission awareness for U.S. and allied military personnel on the ground, in the air, and at sea. They are also the backbone of the blue-force tracking that has greatly reduced casualties from friendly fire, and they underpin the cost- and collateral damage-reducing precision targeting and strike that Americans now expect of kinetic military operations. Although it certainly is possible to communicate, navigate, forecast weather, and participate in global trade without platforms in space, if deprived of them we will not do so as quickly, as well, or with as much precision as is presently the case. Loss or degradation of the information services that space systems provide would not just be inconvenient, but could generate significant hardship and endanger lives.[2]

Chapter 1 outlined how global space activities are rapidly changing in ways that will condition the future of commercial and security activities in and through space. Primary among these is the presumption that continued rapid technological advance and diffusion will bring about a future in which space is a domain in which the United States and many other countries, including our allies and friends as well as potential adversaries, will engage. This rough parity does not mean equivalence in national security capabilities nor will it be even across all space-related functions, but will be most pronounced in areas driven by commerce and globalization, as described in Chapter 1. This reality will inevitably necessitate a paradigm shift from one predicated on the belief that the United States would remain the undisputed leader in space, impervious to challenge by others.

As used in this chapter, "space assets" refer to more than just the satellites in orbit. Without data uplinks to fly and control the satellites and data downlinks to transfer information from space, satellites are of limited use; moreover, much of the data coming from space requires a significant amount of processing to be usable by consumers. Unless otherwise noted, therefore, "space asset" refers to the entire system comprising the physical satellite, data uplink and downlink systems, ground stations, and information processing and distribution. Space assets are characterized as providing three broad categories of information service: commu-

[2] Speaking at a space policy forum, Bruce MacDonald of the United States Institute of Peace offered a useful analogy for thinking about the loss of space to U.S. military capacity as analogous to "being suddenly cut off from gasoline. We would be severely hampered" but would nevertheless remain a powerful force" (Bruce MacDonald, "The New National Space Policy: Prospects for International Cooperation and Making Space Safer for All," remarks a Arms Control Association event, July 1, 2010, Washington, D.C.).

nications; PNT; and observation and surveillance. While the focus of this chapter is on national security space (NSS) assets, the effort focused on safeguarding these specialized assets is critical for civilian systems and users as well.

This chapter explores the requisites for protecting U.S. NSS assets in a domain crowded by state, commercial, and other interests, including transnational criminal and terror groups. It begins with a discussion of scope—namely, the nature of the space domain in U.S. security and how threats to it are characterized. A framework is then provided for evaluating space defense, which is necessary but not sufficient for space security.

THE CHARACTERIZATION OF SPACE IN NATIONAL DISCOURSE

Assessing the necessary national capabilities to defend and protect U.S. NSS assets raises a number of questions. First, what role is space intended to play in U.S. national security? Second, which uses of the space domain does the United States seek to guarantee? Finally, what specifically is the United States attempting to defend the assets from?

The Role of Space in National Security

While many in the NSS community would agree that space is no longer the sparsely populated sanctuary that it once was, debate continues as to what the U.S. national perspective on space as a domain should be and thus what role it should play in national defense. Is it a battlefield to be controlled, or solely a support environment for terrestrial military activity? Does the United States intend to fight in space or through it? Is there any lingering scope for continuing to regard space as somehow different from other geographical regions, and less susceptible to historic trends of terrestrial military competition and conflict? What can the United States do to promote and preserve, as much as possible, the legal regime of outer space as a peaceful environment for all to explore, exploit, and safeguard? More to the point, are offensive or defensive weapons, whether in space or ground-based, a requirement for protecting the ' U.S. or any nation's inherent right of self-defense? Clearly each of these issues has implications for the design, acquisition, and deployment of space systems—activities across the Department of Defense (DoD), the Intelligence Community (IC), and other U.S. government agencies, even in the absence of clear policy guidance. The result is inevitably that design choices are made differently based on the agendas, funding constraints, and other considerations of different organizations, rather than being based on a clearly articulated set of strategic national priorities.

Why has there been a relative dearth of action and policy implementation on this issue? Scholars and practitioners involved in space activities have written extensively on these questions, and successive administrations have recognized the topic. Clear statements on the importance of space systems and the need for what the United States would recognize as "resilience" date back to the Ford administration.[3] There have been numerous high-visibility commissions such as the Rumsfeld Commission of 2001 and the Allard Commission of 2006.[4] There were broad public reactions to the Chinese ASAT test of 2007. More recently, the National Space Policy of 2010 and the National Security Space Strategy of 2011 contain clear policy statements.[5,6] What has not been present is a focus on achieving the stated policy goals, with resources, programs, and people devoted to the task of improving space system protection and defense.

Paradoxically, one reason for this seeming passivity may be the way Americans tend to think about space. It is fair to say that when most Americans think of space or space assets they think largely of the National Aeronautics and Space Administration (NASA), and conceive of space as an open and peaceful expanse for spaceflight, commercial applications, and scientific discovery. In fact, it might be argued that many Americans maintain a somewhat romanticized vision of activities in space, and, in any case, are uncomfortable with the idea of space-based weapons, even for defensive purposes.[7] A national discussion updating public awareness of the changing character of the space domain simply has not yet occurred.[8] In the past, there was in-depth public and professional discussion of the role and purpose of U.S. military power in domains such as the high seas and the air. There is no

---

[3] U.S. Department of State, Office of the Historian, *Foreign Relations of the United States: Volume E-3, Documents on Global Issues, 1973-1976*, released December 18, 2009.

[4] U.S. Department of Defense (DoD), *Report of the Commission to Assess United States National Security Space Management and Organization*, Pursuant to Public Law 106-65, January 11, 2001.

[5] Executive Office of the President, *National Space Policy of the United States of America*, Washington, D.C., June 28, 2010.

[6] DoD and the Office of the Director of National Intelligence (ODNI), *National Security Space Strategy*, Washington, D.C., January 2011.

[7] In a study by the Center for International and Security Studies at Maryland (CISSM), 78 percent of U.S. respondents felt that the United States should not put weapons in space even if these "could serve important military purposes such as protecting" satellites. There was also solid support (78 percent) for the United States engaging in international negotiations to prohibit attacks on another country's satellites even if this was to gain a military advantage, and testing or deployment of ASAT weapons (79 percent) (S. Kull, J. Steinbruner, N. Gallagher, C. Ramsay, and E. Lewis, 2008, *Americans and Russians on Space Weapons: A Joint Study of WorldPublicOpinion.org and the Advanced Methods of Cooperative Security Program, CISSM*, January 24, 2008, http://www.worldpublicopinion.org/pipa/pdf/jan08/CISSM_Space_Jan08.rpt.pdf).

[8] This was the conclusion of a CISSM report published 7 years ago in which researchers, commenting on the Bush Administration's push for "space control," noted that "the American public has not been engaged. In the absence of active negotiations, there has been no prominent congressional discussion of the issues involved. Press coverage of these issues has been very limited." See Kull et al., *Americans and Russians on Space Weapons*, 2008.

obvious reason why space should be exempt, and, as suggested above, there are many compelling reasons to be concerned about threats to space systems.

The discrepancy between the way the American public tends to think of space and the extent of U.S. military and civilian dependence on space systems in daily life creates difficulties for national security and space policy makers and planners tasked with defending these assets. Despite doctrine that speaks to defending and defeating adversaries in space, the observation in a recent article in *Aviation Week* that "many senior officials shy away from using the term 'space control' to describe U.S. national security strategy in space as being too bellicose and sounding too much like militarization of space [for both domestic and international audience consumption]" was corroborated by a number of speakers to the committee.[9] Even the way that missile defense has been framed in public discourse largely erases mention of space weapons.

There are at least two schools of thought on the value of updating U.S. public understanding of space and threats to space assets. One argues that reasoned openness in public discourse would ease barriers among elements of the U.S. bureaucracy working (and funding) space programs. The other argues that increased public discussion of space as a contested domain would be more provocative to U.S. adversaries and to the general public than any value it would provide.

The first school of thought argues that the seemingly broad gap between U.S. public perception and space security requirements is problematic in that it obstructs discussion and analysis of the practicalities of space defense, the costs and trade-offs involved, the implications of the loss of these systems and what to do, defensively and offensively, to protect them. This is not a trivial or solely political issue. It is also not an absolute: updated public education on space matters, and particularly on the types of threats to civilian and military systems that exist, does not mean that all details of U.S. space policy and capabilities are to be made public. Although it is not clear whether classification of space-related issues has led to the public's uninformed view of threats to space assets or vice versa, many of those to whom the committee spoke made it clear that uninformed understanding of space hinders development of national security policy, military concepts of operation, tactics, techniques, and procedures, and the training of the people given the responsibility to defend assets. The lack of understanding can also confound funding and investment decisions as well as assessments of which technologies to develop. This can be a particular challenge for the U.S. military, which seeks to have public support for the policies and actions ordered by its civilian leadership. A lack of clar-

[9] A. Butler, "No sanctuary, Pentagon finally puts up money to defend space assets," *Aviation Week*, May 11-24, 2015. A further irony is that space has arguably been militarized from the beginning but with only sporadic experiments with "weaponization." That may be changing as a result of recent developments.

ity even among experts on the general role that space plays in U.S. security—i.e., the foundational assumptions of space policy—inhibits development of a coherent, multiagency program to defend space systems and activities, and to establish the necessary oversight and political will to see that it is done. An issue with such significant implications for the daily lives and well-being of so many in the United States and abroad should not be derailed by public misconceptions discussed below.

A second school of thought focuses on the risk that broadening the public discussion of space security issues, especially regarding pursuit of offensive "space control" objectives, would send a provocative message to adversaries. This school tends not to believe that the national security community has been unduly hampered in the development of national security space policy by a general pattern of "softer" rhetoric. This community is well accustomed to doing its business with requisite linguistic finesse and sees value in continued reliance upon indirect and diplomatic language in U.S. public discourse about space security. Adherents of this view judge that more frank talk about U.S. national security space objectives will create more opposition, both at home and abroad, which will inhibit development of the space programs the security community seeks to advance. According to this view, public resort to blunter language about space security and offensive or active defensive capabilities would do more to provoke adverse reactions from other space actors than the good it would do in terms of facilitating coordination among U.S. space security and military decision makers. The intelligence and commercial space communities each have their own reasons for public discretion; the one wishes to protect the effectiveness of sources and methods, and the other to minimize barriers to the movement of capital and dual-use technologies in support of global markets.

One point both these schools agree on concerns the adverse consequences of over-classification of materials related to space security and protection. Of course it is essential to protect true national security secrets, but overclassification imposes costs and foregoes important benefits. Secrecy impedes robust professional debate and publication; inhibits public diplomacy; and degrades cross-domain synergies, such as between air and space programs. Unlike other crucial national security activities, such as the protection of submarine capabilities, space protection and defense activities necessarily involve the sharing of information and coordination of action with civil, commercial, and international actors. Steps are being taken to improve information sharing with allies, civil agencies, and select U.S. companies, but the process continues to be slow and difficult.

Space Services: Classifying What Is at Stake

There are numerous ways to categorize the thousands of artificial satellites in orbit today. Often this categorization is done by altitude, orbit, size, or whether they are put to commercial or military use. As technology proceeds, many of these categories are becoming blurred. A visible example is the increased emphasis in national security space policy on military-commercial partnerships (e.g., for commercial supply of some military communications). For the purpose of considering space systems in the national security context, it may be most valuable to classify them by their use—that is, not to consider systems solely in terms of the platform or other technical details, but in terms of the types of information services they provide. This structure also facilitates assessment of the implications to national security of their loss. Following the 2006 RAND *National Security Space Launch Report* schema, the services provided by U.S. satellite systems are categorized into three types most important to military and national security issues: communications; PNT; and Earth observation and surveillance.[10]

Threats to Space Systems and Services

There are many sources of threat to space systems. Naturally occurring threats in space include meteors and fragments, as well as sun flares and other inclement space weather that can damage or destroy the satellite itself or the electronics riding on it. Ground- or air-based components of space systems are also subject to natural causes of service interruption and damage. There are also three categories of human-made threats to assets in space components. For the sake of simplicity the committee will discuss each as a threat to a system. These are (1) collision with space debris—some as small as a paint chip—that can nevertheless disrupt, damage, or destroy the components of a satellite; (2) accidental damage, jamming, or interference; and (3) intentional efforts to damage, degrade, and interfere with or destroy space systems. Threats to ground-based components and supporting infrastructure include, for example, kinetic attack against ground stations and cyber or electronic attack on communications and data networks. Viewed in this way, counter-space threats are not limited to kinetic antisatellite (ASAT) systems, but can occur at multiple nodes of the service-providing system.

Moreover, threats against different parts of these systems can have different effects, different likelihoods, and different mitigation options. While the community of space professionals is keenly aware of the full range of threats to U.S. space- and ground-based systems, there does not appear to be a comprehensive or consistent approach to categorization and risk assessment. Disambiguating different types of

---

[10] Each of these categories can be further subdivided. For example, the RAND National Space Launch Report breaks down communications services into wideband, protected, and narrowband, with a fourth group of data-relay satellites supporting each of these; it breaks down observation satellites into those used for reconnaissance, for missile warning and defense, and for weather monitoring (F. McCartney, P.A. Wilson, L. Bien, T. Hogan, L. Lewis, C. Whitehair, D. Freeman, et al., *National Security Space Launch Report*, MG-503, RAND Corporation, Santa Monica, Calif., 2006, http://www.rand.org/pubs/monographs/MG503.html, p. 2).

threats is an important first step in systematic evaluation of the status of national space assets and the capabilities needed to defend them. Without establishing a comprehensive standard, it becomes difficult to track U.S. government-wide efforts to address threats to space systems, not all of which would necessarily be accomplished by typical space community stakeholders. At present, the basis of evidence for defining a threat can vary greatly, from dated but validated threats to speculative threats based on incomplete but current intelligence.

Self-analysis is a process that is often overlooked in U.S. defense policy development and planning but that has significant implications for our capacity to protect and deter space systems. Particularly in the space domain, a better grasp by DoD, the IC, and policy makers of the international and domestic tolerances for U.S. space activities will be key to augmenting current National Security Space Policy to clarify what the United States—both the security community and the public that ultimately must support its general activities—believes is the proper role of space in national security and the stakes involved in its defense. As discussed below, only once this is done will the United States be able to use implementing directives, declaratory policy, or international agreements to clearly articulate to allies and potential adversaries alike U.S. philosophy about legitimate, safe, and productive uses of outer space, and about the responses to departures from those norms.

DEFENDING AND PROTECTING NATIONAL SECURITY SPACE ASSETS: SPACE DEFENSE TRIAD

The 2011 National Security Space Strategy calls for a "multi-layered approach to prevent and deter aggression" against space systems (Box 2-1).[11]

The security objectives laid out in that strategy suggest a framework of three interrelated means of defending U.S. NSS assets and guaranteeing the national security communication, observation, and PNT services that those assets provide (see Figure 2-1).[12]

---

[11] DoD and ODNI, *National Security Space Strategy*, 2011.

[12] In remarks at the Stimson Center in Washington, D.C., on September 17, 2013, Assistant Secretary of Defense Madelyn Creedon laid out a framework consisting of four mutually supporting components rather than the three described in the present report: "(1) internationalizing norms that enhance stability, (2) building coalitions for collective security, (3) increasing the resilience of our architectures, and (4) being prepared to respond to attacks against U.S. and allied space assets though not necessarily in space." This report combines the first two into *Coalition Formation and International Regimes*, since they both involve discovering common interests, in the first case among allies and friends and in the second more generally and globally. Increasing the resilience of architectures is a subset of *System Protection Measures*, which also includes efforts to improve survivability and other attributes of space systems or deterrence. Finally, the report includes Creedon's fourth component, *Deterrence Measures*. (See Assistant Secretary of Defense Madelyn Creedon, Remarks on

Box 2-1
National Security Space Strategy

The National Security Space Strategy draws upon all elements of national power and requires active U.S. leadership in space. The United States will pursue a set of interrelated strategic approaches to meet our national security space objectives:

- Promote responsible, peaceful, and safe use of space;
- Provide improved U.S. space capabilities;
- Partner with responsible nations, international organizations, and commercial firms;
- Prevent and deter aggression against space infrastructure that supports U.S. national security; and
- Prepare to defeat attacks and to operate in a degraded environment.

SOURCE: U.S. Department of Defense and the Office of the Director of National Intelligence, *National Security Space Strategy*, Washington, D.C., January 2011.

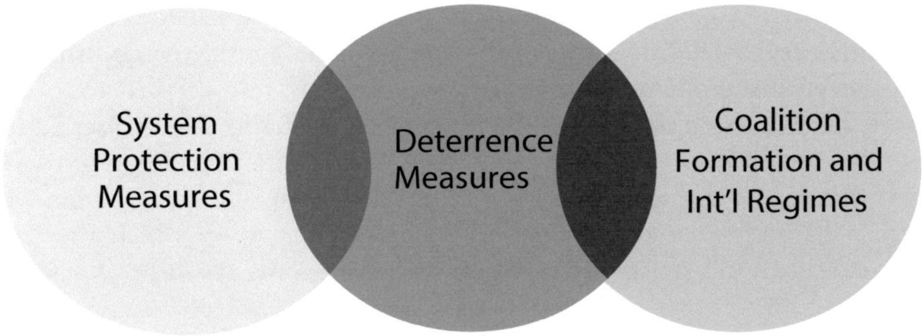

FIGURE 2-1 Elements of defending U.S. national security space assets. SOURCE: U.S. Department of Defense and the Office of the Director of National Intelligence, *National Security Space Strategy*, Washington, D.C., January 2011.

The first element, system protection measures, includes activities that serve the security objectives to prevent and deter aggression and defeat attacks and operate in a degraded environment. These are primarily technological solutions to enhance the survivability of space systems. The second element comprises deterrence messaging measures. The final element of the space defense triad is establishment of coalitions, and international space regimes and norms of behavior that impose

Deterrence, Stimson Center, Washington., D.C., September 2013.) http://www.defense.gov/Portals/1/features/2011/0111_nsss/docs/Stimson-Center-Deterrence-Speech.pdf.)

costs to an adversary—in terms of either having to face a coalition or in loss of diplomatic prestige or other sanctions, to prohibit activities taken against another actor's space systems. Each of these three elements has its own attributes and limitations and no single leg is sufficient for defense; a combination of them is required to ensure a robust defense of NSS assets. Each is discussed in more detail below.

System Protection Measures

The first element of defending space systems involves primarily technical solutions for the purpose of establishing survivability of systems and "deterrence by denial." System protection is the easiest or most straightforward of the three elements, because the United States has the greatest control over what it does and over what opportunities and challenges it presents to its adversaries. Of the three, system protection measures appear to have received the most attention and funding. As one component of a defense triad, system protection involves upgrading current systems where possible, and constructing future systems to be more survivable and thus less vulnerable to collision, interference, or attack. It also involves an array of improvements in system architecture, to make the entire satellite configuration more robust, redundant, and resilient. Improvements in system acquisition and enhancement of interoperability among systems also fit into this category.

Here the discussion highlights how protection relates to the broader issues of defense and deterrence—namely, by establishing the conditions for deterrence by denial by diminishing the probability that an attack against a space system would succeed in degrading or destroying the space services upon which the United States depends. In this sense, protective measures are a main source of deterrence by denial.[13]

Given the number of agencies that have stakes in space, depend on space, and are involved in high-dollar projects to protect those assets, it is critical that these agencies undertake an effort to more fully appreciate their respective interagency authorities and the larger U.S. security environment that will condition national choices and decisions in the event of a space conflict. Any response to an attack

[13] This is consistent with what the DoD Deterrence Operations Joint Operating Concept (DO JOC) refers to as increasing the costs of aggression or denying its benefits. The DoD doctrine defines deterrence as preventing action by the "existence of a credible threat of unacceptable counteraction and/or the belief that the cost of action outweighs the perceived benefits" (DoD, *Dictionary of Military Terms*, 2015, http://www.dtic.mil/doctrine/dod_dictionary/). More specifically, deterrence by denial attempts to forestall an attack or other unwelcome act by persuading an adversary that an attack will not succeed. For example, if the target of the attack is hidden, mobile, or well-shielded, the aggressor may not be able to find it and strike it effectively. Likewise, if the defender is capable of intercepting or disrupting the attack en route, and if the putative attacker is aware of this capability, it may conclude that the attack cannot attain its objective and may thereby be deterred from launching it.

in space will have to take into account the totality of U.S. interests, not just those directly affected by space. This approach can help to better identify and plan against those circumstances in which the country's own processes deny it the full benefit of its capabilities by posing barriers to effective implementation of deterrence messaging or actions. Categorizing and prioritizing risks in space and creating closer whole-of-government response plans are likely to have more value than drawing redlines in space.

Efforts are under way to organize DoD space efforts as a formal major force program (MFP-12) that would provide a more integrated understanding of the resources being devoted to all defense space activities. This action should better enable the DoD to set priorities and assign resources for the protection and defense of its space assets. Doing so should also enable DoD to better coordinate with the IC, U.S. industry, and allies in protecting and defending non-DoD space assets crucial to U.S. national security. Such efforts will require more than purely technical or military capabilities, but economic, diplomatic, and political arrangements that align the interests of other space–actor nations with those of the United States. In addition to the traditional forms of deterrence, such as denial of attack objectives and fear of retaliation, U.S. security can be improved by leading space-related cooperative efforts that reduce or redirect incentives to come into conflict with the United States.

Deterrence Measures

The second leg of the space defense triad is deterrence: discouraging people, groups, or states from interfering with or attacking U.S. space systems either because there is no value in doing so or the actual or threatened cost of doing so is too high.[14] It is served by both system protection measures and participation in the development of international regimes governing space. Importantly, the success of deterrence measures against attacks on U.S. space systems, like other forms of deterrence, is in the perception of the would-be attacker. It is a political and psychological strategy "that must be directed by political leaders, coordinated with diplomacy, and sensitive to the adversary's political constraints, world views, and

[14] Deterrence by threat of retaliation was at the core of the U.S.-U.S.S.R. nuclear relationship during the Cold War. Each side was deterred from launching a first strike against the other for fear of unacceptable retaliation. Much strategic thinking and weapons systems development was devoted to ensuring that, under any combination of circumstances, the ability to launch a devastating second strike was preserved, so that neither side could calculate that a first strike would be profitable. Notably, the threatened response to a hostile action need not be in kind—retaliation in a different time, place, and manner may be more effective in influencing the calculation of the adversary, because the responder can exercise freedom to reply in a manner tailored to its advantage.

perceptions."[15] As such it is not an outcome that can be induced unilaterally, but the result of an exchange or interaction between a deterrer and a deterree. As Milevski puts it, "One cannot pull deterrence out of a toolbox and employ it. It must be induced in the other." It is the result of an adversary's choice to be deterred.[16]

Effective deterrence is classically based on three principles: (1) credibility, often associated with a state's perceived resolve or political will that a threatened response can and will be executed if a red line is crossed; (2) possession of coercive capability sufficient and appropriate to hold an adversary's valued assets at risk, and to implement a threatened response to an unwanted action; and (3) the ability to communicate to a potential adversary what actions are to be avoided and the nature of the intended response (punishment). In each of these elements, the United States must draw upon the full array of elements of national power, including diplomatic, intelligence, military, and economic tools.

Credibility of a Deterrent Threat

The United States' ability to deter peer, near-peer, and non-state actor efforts to interfere with or attack space systems requires that deterrent threats be credible. First, a potential attacker must believe that the defender can identify the source of an attack should one occur. An adversary that does not fear being caught has much greater latitude for action. This requires that potential adversaries believe that the United States maintains the capacity to detect, track, and identify a full range of space objects and to distinguish hostile attack from system failures, space weather, or other natural phenomena. Deterrence efforts can be rendered effectively useless by an inability to recognize whether a system has been struck or interfered with, where in the system the attack has occurred, and who did it. Under those conditions an adversary could be incentivized to strike first in any conflict (military or otherwise), especially since an effective attack on the enemy's satellites may well degrade its ability to respond in any coordinated and effective manner. It is important to note that the issue is not just one of diagnosis and attribution however; it is timely attribution. Adversaries considering interference with U.S. space systems need not believe that they would never be found out in order for deterrence to fail, just that the interference remains undetected long enough to achieve the desired operational, tactical, or political objective.

---

[15] J. Levy, Deterrence and coercive diplomacy: The contributions of Alexander George, *Political Psychology* 29(4):537-552, 2008.

[16] L. Milevski, Deterring "Able Archer:" Comments arising from Adamsky's "Lessons for Deterrence Theory and Practice," *Journal of Strategic Studies* 37(6-7):1050-1065, 2014, doi:10.1080/01402390.2014.952408.

Potential adversaries contemplating interference with U.S. space systems should expect the United States to execute any necessary and proportional response, but not necessarily in space. In considering the totality of U.S. interests in any particular conflict, the protection and defense of U.S. and allied space assets may be considered as part of continuously updated war plans of each combatant commander in their respective areas of responsibility, in coordination with the U.S. Strategic Command. Routine exercises could include a wide range of retaliatory actions across different domains to enhance the credibility of U.S. diplomatic and declaratory statements, through public and nonpublic channels.

Capability of Responding

Because they are so crucial to the credibility of deterrent messages, rapid and accurate diagnosis of an attack wherever it might occur across the entirety of a space system and identification of its source are among the most essential capabilities for space defense and protection. They are appropriately prioritized in U.S. National Space Policy and directives regarding space situational awareness (SSA). What is not clear is the extent to which the priority for enhanced awareness is also applied to terrestrial aspects of space systems. Finally, it is important that these efforts are managed carefully with progress milestones and effectiveness measurement. This should occur not only on the technical side. SSA requirements and capability should not just feature space security policy but should also find their place in national security policy and implementing plans.

Additional requirements for the credibility of deterrence messages are that potential adversaries believe that the deterrer has both the capability and the will to respond to an attack once identified. It is not difficult to argue that U.S. defense forces could respond forcefully to deterrence failure in any number of ways. The United States can respond. However, the credibility conundrum, often cited with regard to nuclear weapons, is the believability that the United States will respond in circumstances short of catastrophic damage or loss of conventional capacity. The nature and scale if not the details of response to all-out, strategic attack on the homeland are fairly easy to imagine. However, if that attack were to avoid immediate loss of life, for example, what would be an appropriate level of response? Is limited attack that disables mission critical space service in an area of operations during a crisis more or less escalatory than total destruction of a global observation system in peacetime? Given that the fundamental value of almost all current U.S. space systems comes from the information they provide, there are potentially useful analogies to be drawn with cyber-related attacks and responses. The purpose of an intentional attack is not merely to disrupt or destroy the function of a space or cyber capability, but by doing so, to achieve some political or military end. Figure 2-2 is derived from a RAND report on responding to cyberattacks of various kinds, including the problem of attribution.

It is important to stress here the possibilities for cross-domain deterrence. That is, a threat to a space system need not be countered exclusively, or even mainly, in space. In appropriate circumstances, a response targeting an enemy's ground, sea or air systems might be more effective than retaliation in kind. Especially when the United States is so much more invested in space—receiving so much benefit from its satellites and accordingly becoming so much more dependent upon them—artificially confining a confrontation to space may work sharply to our disadvantage. It is critical to maintain flexibility in our deterrent posture, to be able to respond to a threat to NSS assets at a time, place, and manner of our choosing.

In responding to attacks on cyber and space systems, it is helpful to understand the many sources of uncertainty for adversaries in mounting an attack and for the United States in responding to one. At a technical level, there can be difference between what the attack intended and what it achieved (what we term here as "error"). At the assessment level, there can be differences between the actual effect of the attack and how the attack is perceived by both the attacker and the defender (what we term here as "misperception"). Misperception can be due to both technical and nontechnical issues, such as inadequate SSA or poor political understanding of an adversary's motives. Finally, in responding to an attack, whether in a symmetrical or asymmetrical manner, the actual and perceived effects of the responses may be assessed differently by the adversaries (what is termed "miscalculation). There may be a technical aspect to miscalculations, such as inadequate intelligence or understanding of an adversary's vulnerabilities, but the primary challenge will be lack of understanding of an adversary's motives and values—an understanding that is also necessary to the effective creation of deterrence in the adversary's mind. Reducing the sources of error in the decision cycle for responding to space attacks requires capabilities that also improve the resilience of space systems. Some of these needed capabilities are technical (e.g., models of adversary satellites and communications systems) while others require better insight into adversary thinking and a better understanding of what is important to the United States and its allies. The last point goes beyond an understanding of an adversary's doctrine, organization, and war plans, but calls for insight into what the adversary desires and fears. This allows for a better understanding of an adversary's internal political calculations, so as to better tailor U.S. deterrence capabilities and actions.

Communicating Deterrence Messages

What is the deterrence objective with regard to space assets that the United States seeks to communicate? U.S. policy statements, and indeed strategy, are still developing and at present remain somewhat ambiguous on this point, although

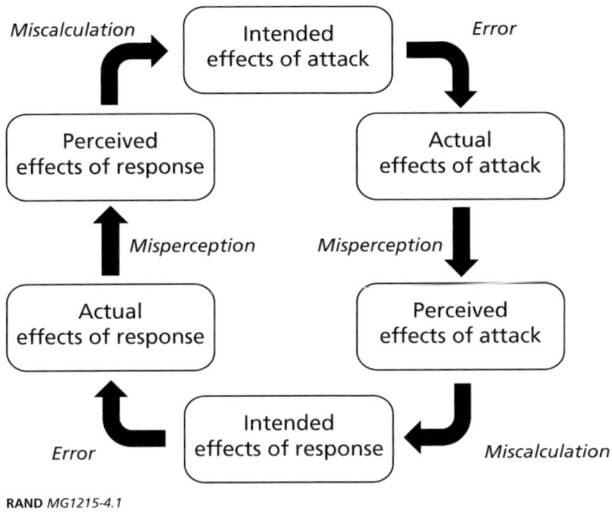

FIGURE 2-2 Sources of imprecision in responding to an attack. SOURCE: Derived from M.C. Libicki, *Crisis and Escalation in Cyberspace*, MG1215-4.1, RAND Corporation, http://www.rand.org/pubs/monographs/MG1215.html. Reprinted with permission from the RAND Corporation, Santa Monica, California.

sometimes a degree of studied ambiguity can be a useful component of a deterrence strategy. It is relatively clear that the broad goal is to deter others from attempting to interfere with, deny, degrade or destroy the space services (e.g., satellite-based communication, satellite-based sensors, and positioning, navigation, and timing) that the United States uses during peacetime, crisis, and war. The committee has described what the objective and threat might be. Deciding on and communicating response options also requires a good understanding of why a potential adversary might act. Classical deterrence theory often takes the "why" out of consideration because it assumes both aggression and the means of mitigation—namely, when rule violations and aggression occur, costs can be lower than gains when actors do not fear retaliation. Applied in the security context of the Cold War, costs were generally taken to be the loss of some military capacity or valued target by kinetic means. In a few instances costs could also be induced by international sanctions for violations of established norms. However, as evidenced by U.S. counterterror efforts over nearly two decades, eliciting a destructive or violent response for other than battlefield effect can be an adversary's objective. The point is, it is essential for national security decision makers to consider the "why" of adversary behavior when deterrent threats and possible responses to deterrence failure are developed.

This is the impetus behind 'a pivotal article on tailoring U.S. deterrence activities to 21st century adversaries.[17] It is no less the case in the space domain.

To date, most discussion of which actors might threaten NSS assets and thus to whom U.S. deterrent messages should be accessible focuses on states. In particular, China and Russia might view the ability to deny the United States use of space as an important means of deterring U.S. conventional military operations, and as a force multiplier in the case that their deterrence efforts fail and they finds themselves in a military conflict with the United States. However, looking forward, the list of potential adversaries with incentives to degrade or destroy U.S. NSS assets does not, and will not, end with China and Russia. As a recent *Foreign Affairs* article points out, with rapid advances in microcomputing as well as lower costs of commercial launch and ready-made satellites, new types of actors will without doubt be entering the space-faring community.[18] While most will be research- or commerce-oriented, it is also the case that many more than those we now consider space threats will soon enter the fray. As for threats to U.S. national security, these could include non-state actors with aggressive intentions; criminal organizations; and even the commercial entities that U.S. policy makers look to in order to stem the cost of a fully government-funded space program. As technology progresses, proliferates and inevitably becomes less costly, a critical feature of the defense strategies possessed by even non-space-faring actors will include denying the use of space to militarily superior foes, so as to remove space as a source of strategic advantage.

Regarding some of the smaller space powers, the two types of deterrence discussed above will still be operating, but perhaps with somewhat different effect. For example, deterrence by denial might be somewhat easier, if emerging space players will at least initially be capable of launching only relatively less sophisticated (and less numerous) attacks on U.S. satellites; however, even unsophisticated attacks might be quite devastating to an insufficiently resilient satellite system. Deterrence by threat of U.S. retaliation would likely continue to be robust, as the United States would likely have multiple avenues for counterattacking the aggressor state, but it would likely need to be an asymmetric attack, because the new space powers would probably not be so dependent upon their own satellite services that U.S. disruption would be equally important to them.

[17] M.E. Bunn, "Can Deterrence Be Tailored?" Strategic Forum, National Defense University, No. 225, January 2007.

[18] D. Baiocchi and W. Welser IV, The Democratization of Space, *Foreign Affairs*, May-June 2015.

Coalition Formation and International Regimes

The third leg of the deterrence triad focuses on the advantages of enlisting additional participants and the possibility of enhancing the security of U.S. space programs and activities by leveraging international coalitions and regimes. As noted, the United States space security actors enjoy partnerships with a wide array of players: commercial operations in the United States and elsewhere; multiple friendly states with substantial interests in peaceful operations in space that parallel our own; nongovernmental organizations, and others. These diverse capabilities support deterrence by providing multiple redundant pathways for the performance of vital space services, reducing the vulnerability of the United States. By generating less dependency upon any single space vehicle or network, these coalitions can diminish an adversary's expectations that an attack on any one, or a few, satellite nodes could effectively deny space services.

International actors can also contribute to deterrence by raising the political price of hostile space actions, by giving more players a direct stake in avoiding hostilities, and by creating more opportunity for pressuring the hostile actors to avoid arousing the entire community. International law, through the creation and advancement of legally binding treaties and other meaningful norms of behavior, can contribute to this form of deterrence as well. No new agreements of this sort have been concluded for several decades, and none seems immediately on the horizon. The United States has not taken advantage of the opportunity to lead the world in the consideration of additional possibly useful norms here.

Outer space, like other areas of international relations, has seen its share of arms racing. Many factors contribute to a country's decisions about pursuing space weapons, but one of them undoubtedly is the actions of its erstwhile rivals. In an arms race, it can be unclear who started it and it who is ahead may also be debatable. Self-restraint at any point by any one country, including the United States, may not be effective in dampening down an emerging space arms race. But it does seem likely that if the United States pioneers a new development in space weaponry, other actors are likely to follow, more or less, sooner or later. However, since the United States depends on its NSS assets more than any other country, it has a big stake in promoting the security and sustainability of the space environment, and thus a strong interest in avoiding a space arms race—even if it were an arms race that the United States could, in some sense, win owing to its technological advantages.

In responding to the recent space weapons activities of other countries, therefore, the United States must be attentive to the long run—not just whether it can effectively outdo the most recent actions of China and Russia, but how those states will later counterrespond to our moves, and what cycle of competition may ensue. The premium, accordingly, should be on measures to negate or mitigate, to the extent possible, the weapons advances made by Russia and China, without stimulating them to proceed even further and faster in a direction that is distinctly disadvantageous to the United States.

Exercises that include groups from across the DoD, IC and combatant commands as well as participation from the Department of State and other interested space stakeholders offer both practical training and opportunities for deterrence messaging. Including domestic security and public resilience services in the simulation of emergency situations enhances these opportunities. Exercising the loss of systems including cross-service, emergency management and response, and public resiliency is one way to enhance the U.S. deterrence message by reducing the possible effect of space attacks. U.S. space policy cannot intelligently be made without regard for the policies and practices of other states, and without attention to the likely interaction between our choices and theirs. Indeed, the United States does not unilaterally decide questions about the future security of outer space: We surely have a voice—arguably the most important single voice—on those matters, but many stakeholders will participate, and will respond to our words and their own interpretations of our actions. Finally, the fact that the United States is unlikely to be fighting alone against peer or near-peer adversaries is important when it considers appropriate space security strategies. The United States is inextricably linked and dependent upon its allies to fight with it—something that was well recognized by the establishment of interoperability standards—for example, the North Atlantic Treaty Organization standardization agreements (STANAGS) and common field training venues. Extending this paradigm to the space domain is critical for overall net resilience.

FINAL THOUGHTS

The purpose of this chapter has been to discuss the requisites for the nonmaterial aspects of the defense and protection of U.S. NSS assets. The importance of these assets to the safety and quality of life for people around the world and here at home is commonly underappreciated. However, there exists an opportunity today for the relatively small NSS community to engage the Congress and the public in a national discussion about the threat to U.S. NSS assets, the U.S. role in space, and the types of activities the United States should be engaging in. Opening up the public discussion on what has fast become the new normal in space—an environment that is crowded with debris and one that international actors may seek to use for aggressive purposes—would educate the public about the threats to aspects of daily life often taken for granted. Within the limits of security classification and with sensitivity for how the discussion may be received abroad, a new openness could also facilitate discussion of what it will take in the current circumstances to defend U.S. and global space assets from man-made efforts to degrade or defeat them.

Appendixes

A

Biographical Sketches of Committee Members

JAMES O. ELLIS, JR., *Co-Chair*, currently serves as an Annenberg Distinguished Visiting Fellow at the Hoover Institution at Stanford University. He retired as president and chief executive officer of the Institute of Nuclear Power Operations (INPO), in Atlanta, Georgia, in 2012. In 2004, Admiral Ellis completed a distinguished 39-year Navy career as commander of the United States Strategic Command during a time of challenge and change. In this role, he was responsible for the global command and control of U.S. strategic and space forces, reporting directly to the Secretary of Defense. A 1969 graduate of the U.S. Naval Academy, Admiral Ellis was designated a Naval aviator in 1971. His service as a Navy fighter pilot included tours with two fighter squadrons and assignment as commanding officer of an F/A-18 strike/fighter squadron. In 1991, he assumed command of the USS Abraham Lincoln, a nuclear-powered aircraft carrier. After selection to rear admiral, in 1996 he served as a carrier battle group commander leading contingency response operations in the Taiwan Straits. His shore assignments included senior military staff tours directing operations for the U.S. Atlantic Fleet and as Deputy Chief of Naval Operations (Plans, Policy, and Operations). He also served as commander in chief, U.S. Naval Forces, Europe, and commander in chief, Allied Forces, Southern Europe, during a time of historic NATO expansion and led United States and NATO forces in combat and humanitarian operations during the 1999 Kosovo crisis. Mr. Ellis holds an M.S. in aerospace engineering from the Georgia Institute of Technology and, in 2005, was inducted into the school's Engineering Hall of Fame. He also has an M.S. in aeronautical systems from the University of West Florida. He completed U.S. Navy nuclear power training and was qualified in the operation and maintenance of naval nuclear propulsion plants. He is a graduate of the Navy Test Pilot School, the Navy Fighter Weapons School (Top Gun), and the Senior Officer Program in National Security Strategy at Harvard University. In 2013, Mr. Ellis was elected to the National Academy of Engineering. His personal awards include the Defense Distinguished Service Medal (three awards), Navy Distinguished Service Medal (two awards), Legion of Merit (four awards), Defense Meritorious Service Medal, Meritorious Service Medal (two awards), and the Navy Commendation Medal, as well as numerous campaign and service awards. He was presented the Order of Merit of the Republic of Hungary, the Star of Merit and Honor from the Greek Ministry of Defense, the Joint Forces Medal of Honor, and the Grand Order of Merit of the Italian Republic. Mr. Ellis currently serves on the board of directors of the Lockheed Martin Corporation, Dominion Resources, and Level 3 Communications, where he is the non-executive chairman of the board. In 2009, he completed 3 years of service as a Presidential appointee on the President's Intelligence Advisory Board, and in 2006 he was a member of the Military Advisory Panel to the Iraq Study Group.

MARTIN C. FAGA, *Co-Chair*, is a retired president and chief executive officer of the MITRE Corporation. He was a member of the MITRE Board of Trustees until 2012. Before joining MITRE, Mr. Faga served from 1989 until 1993 as Assistant Secretary of the Air Force for Space, where he was responsible for overall supervision of Air Force space matters. At the same time, he served as director of the National Reconnaissance Office (NRO), responsible to the Secretary of Defense and the Director of Central Intelligence for the development, acquisition, and operation of all U.S. satellite reconnaissance programs. Mr. Faga is a fellow of the National Academy of Public Administration and is a member of the board of directors of the Association of Former Intelligence Officers. He served from 2006 until 2009 on the President's Intelligence Advisory Board and was a member of the Public Interest Declassification Board from 2006 to 2014. Since retiring from MITRE, Mr. Faga has been elected to the boards of directors of Orbital ATK, DigitalGlobe, and Inmarsat Government. He is chairman of the board of Thomson Reuters Special Services. He has also served on the board of Electronic Data Systems. Mr. Faga received M.S. and B.S. degrees in electrical engineering from Lehigh University in 1964 and 1963.

ALLISON ASTORINO-COURTOIS is executive vice president at National Security Innovations (NSI), Inc. She has served as technical lead on a number of multilayer analysis projects sponsored by the Office of the Secretary of Defense in support of U.S. forces and combatant commands. Prior to joining NSI, Dr. Astorino-Courtois worked for Science Applications International Corporation (2004-2007), where she served as a U.S. Strategic Command liaison to U.S. and international communities, and was a tenured associate professor of international relations at Texas

A&M University (1994-2003), where her research focused on the cognitive aspects of foreign policy decision making. She has received a number of academic grants and awards and has published articles in multiple peer-reviewed journals including *International Studies Quarterly, Journal of Conflict Resolution, Political Psychology, Journal of Politics,* and *Conflict Management and Peace Science.* She has also taught at Creighton University and as a visiting instructor at the U.S. Military Academy at West Point. Dr. Astorino-Courtois earned her Ph.D. in international relations from New York University. She served as a co-chair of the Committee on U.S. Air Force Strategic Deterrence Military Capabilities in the 21st Century Security Environment of the National Academies of Sciences, Engineering, and Medicine.

OWEN C. BROWN is a solutions architect with SAIC, following his role as chief technology officer of Kinsay Technical Services, Inc. (KTSi). In that role he acted as the lead executive responsible for management, development, and integration of the company's intellectual offering, enabling and enhancing customer technical objectives. He provides direct support to the Defense Advanced Research Projects Agency (DARPA) and the U.S. Air Force on a variety of complex space system programs. His efforts include assessments of ongoing programs directly to the commander of Air Force Space Command. From 2003 to 2009 Dr. Brown was a program manager in DARPA's Tactical Technology Office, where he managed multiple small spacecraft programs. He led the MiTEx space program from design to highly successful on-orbit demonstration and was later recognized as one of DARPA's top program managers for this effort. While at DARPA Dr. Brown created the fractionated spacecraft architectural concept and led the initial stages of the associated System F6 program. He worked for several years as a spacecraft engineer at Space Systems/Loral supporting the design, test, build, and launch of multiple geosynchronous spacecraft for customers including Intelsat, NTT DoCoMo, NASA/NOAA, DishTV, and the International Space Station. Dr. Brown served as a nuclear submarine officer onboard fast attack submarines and retired after completing 20 combined years of active duty and reserve service. He holds an M.S. and a Ph.D. in aeronautical and astronautical engineering from Stanford University.

VINCENT W.S. CHAN is the Joan and Irwin Jacobs Chair Professor of electrical engineering and computer science at the Massachusetts Institute of Technology (MIT). From 1974 to 1977, he was an assistant professor of electrical engineering at Cornell University. He joined MIT Lincoln Laboratory in 1977 and has been division head of the Communications and Information Technology Division until becoming the director of the Laboratory for Information and Decision Systems (1999-2007) at MIT. He founded and is currently a member of the Claude E. Shannon Communication and Network Group at MIT's Research Laboratory of Electronics. In July 1983, he initiated the Laser Intersatellite Transmission Experiment Program and in 1997, the follow-on GeoLITE Program. In 1989, he led the All-Optical-Network Consortium (1990-1997) formed among MIT, AT&T, and the Digital Equipment Corporation. He also served as principal investigator of the Next Generation Internet Consortium, ONRAMP (1998-2003), formed by AT&T, Cabletron, MIT, Nortel, and JDS; and the Satellite Networking Research Consortium funded by the National Science Foundation and formed between MIT, Motorola, Teledesic, and Globalstar. In 2009, he founded and served as the editor-in-chief of the *Journal of Optical Communications and Networking* until 2012. He has served many government advisory boards and is currently a member of the Corporation of Draper Laboratory and was on the board of governors of the Institute of Electrical and Electronics Engineers (IEEE) Communication Society as vice president of publications. He is currently serving on the National Security Agency Advisory Board's research and technology panels and was on the most recent Intelligence Science Board of the Office of the Director of National Intelligence. He is an elected member of Eta Kappa Nu, Tau Beta Pi, and Sigma Xi, and a fellow of IEEE and of the Optical Society of America. Throughout his career, Dr. Chan has focused his research on communication and networks, particularly on free space and fiber-optical communication and networks and satellite communications. His work has led the way to the first successful ultra-high-rate laser communication demonstration in space and early deployment of WDM optical networks. His recent research emphasis is on heterogeneous (SATCOM, wireless, and fiber) network architectures with stringent performance demands. He received his Ph.D. in electrical engineering from MIT.

MICHAEL D. GRIFFIN is the chairman and chief executive officer of Schafer Corporation, a leading provider of scientific, engineering, and technical services and products in the national security sector. He was previously King-McDonald Eminent Scholar and Professor of Mechanical and Aerospace Engineering at the University of Alabama, Huntsville, was the Administrator of NASA from 2005 to 2009, and prior to that was the Space Department head at the Johns Hopkins University Applied Physics Laboratory. He has also held numerous executive positions with industry, including president and chief operating officer of In-Q-Tel, chief executive officer of Magellan Systems, general manager of Orbital Science Corporation's Space Systems Group, and executive vice president and chief technical officer at Orbital. Dr. Griffin's earlier career included service as both chief engineer and associate administrator for exploration at NASA, and as the deputy for technology at the Strategic Defense Initiative Organization (SDIO). Prior to joining SDIO in an executive capacity, he played a key role in conceiving and directing several first-of-a-kind space tests in support of strategic defense research, development, and flight testing. These included the first space-to-space intercept of a ballistic missile in powered flight, the first broad-spectrum spaceborne reconnaissance of targets and decoys in midcourse flight, and the first space-to-ground reconnaissance of ballistic missiles during the boost phase. He also played a leading role in other space missions in earlier work at the JHU Applied Physics Laboratory, NASA's Jet Propulsion Laboratory, and the Computer Science Corporation. Dr. Griffin was an adjunct professor for 13 years at the University of Maryland, Johns Hopkins University, and George Washington University, teaching courses in spacecraft design, applied mathematics, guidance and navigation, compressible flow, computational fluid dynamics, spacecraft attitude control, astrodynamics, and introductory aerospace engineering. He is a registered professional engineer in Maryland and California and is the lead author of over two dozen technical papers and the textbook *Space Vehicle Design*. Dr. Griffin is a member of the National Academy of Engineering and the International Academy of Astronautics, an honorary fellow and the current president of the American Institute of Aeronautics and Astronautics (AIAA), a fellow of the American Astronautical Society, and a senior member of IEEE. He is the recipient of numerous honors and awards, including the NASA Exceptional Achievement Medal, the AIAA Space Systems Medal and Goddard Astronautics Award, the National Space Club's Goddard Trophy, the Rotary National Award for Space Achievement, the Missile Defense Agency's Ronald Reagan Award, and the Department of Defense (DoD) Distinguished Public Service Medal, the highest award that can be conferred on a nongovernment employee. He received his Ph.D. in aerospace engineering from the University of Maryland and has been recognized with honorary doctoral degrees from Florida Southern College and the University of Notre Dame.

RAYMOND JEANLOZ is professor in Earth and planetary science and in astronomy, and is senior fellow in the Miller Institute for Basic Research in Science at the University of California, Berkeley. His specialties include the constitution and evolution of planetary interiors and properties of materials at high pressures and temperatures. After completing his Ph.D. at the California Institute of Technology, he was on the faculty of Harvard University and then moved to UC Berkeley. Dr. Jeanloz has served as an advisor to academia, industry, and government, including as chair of the Academies' Board on Earth Sciences and Resources (2000 to 2002) and of the NAS Committee on International Security and Arms Control (since 2005). Dr. Jeanloz is a member of the National Academy of Sciences and a member of the Science and Technology Committee advising the LLCs that manage Los Alamos and Livermore laboratories. He is a fellow of the American Academy of Arts and Sciences, the American Association for the Advancement of Science, the American Geophysical Union, and the American Physical Society.

DAVID A. KOPLOW is a professor of law at Georgetown University. He specializes in the areas of public international law and national security law. Professor Koplow joined the Georgetown law faculty in 1981. His principal courses have been International Law I (the introductory survey of public international law topics), a seminar in the area of arms control, nonproliferation and terrorism, and the proseminar for LLM students in national security law. In addition, he has directed a clinic, the Center for Applied Legal Studies, in which students provide pro bono representation to refugees who seek asylum in the United States because of persecution in their homelands. His government service has included stints as special counsel for arms control to the general counsel of DoD (2009-2011); as deputy general counsel for international affairs at DoD (1997-1999); and as attorney-advisor and special assistant to the director of the U.S. Arms Control and Disarmament Agency (1978-1981). He is a graduate of Harvard College and Yale Law School and was a Rhodes Scholar. Most of his scholarly writing concentrates on the intersection of international law and U.S. constitutional law, especially in the areas of arms control and national security and treaty negotiation and implementation. He received a J.D. from Yale University.

L. ROGER MASON, JR., is senior vice president, national security and intelligence, and chief security officer at Noblis. Dr. Mason serves as senior vice president and corporate officer responsible for the overall direction of Noblis' national security missions, including intelligence, defense, homeland security, and law enforcement. He returns to Noblis after 5 years of service in the Intelligence Community (IC) as the first assistant director of national intelligence for Systems and Resource Analyses (ADNI/SRA). In this capacity, Dr. Mason served as the DNI's principal intelligence officer and trusted advisor on all matters dealing with intelligence capabilities, resources, requirements, systems analysis, program evaluation, and cost analysis. He led the establishment of this new capability that combined operations research, decision sciences, and business analytics to aid the DNI and senior intelligence agency leaders make difficult decisions on complex issues that spanned every aspect of intelligence from overhead space technologies to counterterrorism. In recognition of his service, Dr. Mason was awarded the National Intelligence Distinguished Service Medal—the IC's highest award. In addition, he led SRA to four National Intelligence Meritorious Unit commendations and received numerous intelligence agency awards. Prior to federal service, Dr. Mason served in a number of senior executive positions in the national security sector, including vice president at Noblis, director at the Institute for Defense Analyses, and general manager of the Advanced Systems Group at General Dynamics (formerly Veridian). Earlier in his career, he led a number of advanced programs combining technology development, system integration, and field operations for military and intelligence missions. He is a nationally recognized expert in intelligence capabilities, operations research, overhead reconnaissance, systems integration, and change leadership. He has published more than 35 papers in peer-reviewed journals and symposia and holds two

U.S. patents dealing with advanced materials and collection devices. Dr. Mason earned his Ph.D. and M.S. in engineering physics (nuclear) from the University of Virginia, a master's degree in business administration from the Northwestern University's Kellogg School, and a B.S. in physics from George Washington University. Additionally, he has been recognized with many professional awards, including the Omicron Delta Kappa Leadership Honor Society, Alpha Nu Sigma Honor Society for nuclear science, and the University of Virginia Distinguished Student Award. Dr. Mason is an active leader in the Boy Scouts of America and has been a part of this organization for over 37 years, including attaining the rank of Gold Palm Eagle Scout.

JOHN A. MONTGOMERY is the director of research at the Naval Research Laboratory, where he oversees research and development programs with expenditures of approximately $1.2 billion per year. He joined the Naval Research Laboratory (NRL) in 1968 as a research physicist in the Advanced Techniques Branch of the Electronic Warfare Division, where he conducted research on a wide range of electronic warfare (EW) topics. In 1980, he was selected to head the Off-Board Countermeasures Branch. In May 1985, he was appointed to the Senior Executive Service (SES) and was selected as superintendent of the Tactical EW Division. He has been responsible for numerous systems that have been developed/approved for operational use by the Navy and other services. He has had great impact through the application of advanced technologies to solve unusual or severe operational deficiencies noted during world crises, most recently in Afghanistan and Iraq and for Homeland Defense and in the Pacific theater. Dr. Montgomery received the Department of Defense Distinguished Civilian Service Award in 2001. He was recognized by the Department of the Navy Distinguished Civilian Service Award in 1999 and by the Department of the Navy Meritorious Civilian Service Award in 1986. As a member of the SES, he received the Presidential Rank Award of Distinguished Executive in 1991 and again in 2002, and the Presidential Rank Award of Meritorious Executive in 1988, 1999, and again in 2007. He also received the 1997 Dr. Arthur E. Bisson Prize for Naval Technology Achievement, awarded by the Chief of Naval Research in 1998. Further, he received the Association of Old Crows (Electronic Defense Association) Joint Services Award in 1993. He was an NRL Edison Scholar and is a member of Sigma Xi. He served as the U.S. National Leader of the Technical Cooperation Program's multinational group on EW from 1987 to 2002, and served as its executive chairman. In 2006, Dr. Montgomery received the Laboratory Director of the Year Award from the Federal Laboratory Consortium for Technology Transfer, and in 2011, he received the Roger W. Jones Award for Executive Leadership from American University's School of Public Affairs. Dr. Montgomery received his Ph.D. in physics from the Catholic University of America. He is a member of the National Academy of Engineering.

SCOTT PACE is the director of the Space Policy Institute and professor of the practice of international affairs at the George Washington University's Elliott School of International Affairs. His research interests include civil, commercial, and national security space policy, and the management of technical innovation. From 2005 to 2008, he served as the associate administrator for program analysis and evaluation at NASA. In this capacity, he was responsible for providing objective studies and analyses in support of policy, program, and budget decisions by the NASA Administrator. He previously served as chief technologist for space communications in NASA's Office of Space Operations, where he was responsible for issues related to space-based information systems. He participated in negotiations that resulted in the 2004 GPS-Galileo Agreement between the United States and the European Commission. Dr. Pace also previously served as the deputy chief of staff to NASA Administrator Sean O'Keefe. His primary areas of responsibility included oversight of the President's management agenda in human capital, competitive sourcing, expanding e-government, financial management, and integrating budget and performance. Prior to NASA, Dr. Pace was the assistant director for space and aeronautics in the White House Office of Science and Technology Policy (OSTP). There he was responsible for space and aviation-related issues and coordination of civil and commercial space issues through the Space Policy Coordinating Committee of the National Security Council. From 1993 to 2000, Dr. Pace worked for the RAND Corporation's Science and Technology Policy Institute—a federally funded research and development center for OSTP. Dr. Pace was a key member of a successful international effort to preserve radio navigation satellite spectrum at the 1997 World Radiocommunication Conference (WRC-97) and the addition of new spectrum for satellite navigation at WRC-2000. He also was a member of the DoD Senior Review Group on Commercial Remote Sensing and the Academies' Committee on Earth Sciences. From 1990 to 1993, Dr. Pace served as the deputy director and acting director of the Office of Space Commerce, in the Office of the Deputy Secretary of the Department of Commerce. Dr. Pace represented the department to the National Space Council and participated in efforts affecting export controls for space technologies, space trade negotiations with Japan, Russia, China, and Europe, the licensing process for private remote sensing systems, missile proliferation, and the U.S. space industrial base. Dr. Pace received a Ph.D. in policy analysis from the RAND Graduate School.

THOMAS E. ROMESSER is an independent consultant. Dr. Romesser was chief technology officer for Northrop Grumman Aerospace Systems until the start of 2012 and sector vice president of Aerospace Systems. In those roles, he provided senior leadership representation with customers, universities, industry, and the rest of the corporation. He also was responsible for technology development to support future programs while maintaining close linkage to legacy programs. Prior to his present assignment, Dr. Romesser was sector vice president and general manager of the Technology and Emerging Systems Division for Northrop Grumman's former Space Technology sector. In this role, he was responsible for the development and execution of Space Technology's strategy to support both near- and long-term business objectives, system enhancements and technology leverage for new business pursuits. He oversaw activities of the Directed Energy Systems and Advanced Concepts organizations as well as the Space Technology Research Laboratories. Previously, Dr. Romesser was vice president of technology development; responsible for the identification, development, and acquisition of Space Technology's strategic technologies; and managed discretionary investments in technology and product development. He joined Northrop Grumman via the acquisition of TRW in 2002. A vice president since 1998, he previously served as vice president and deputy of the Space and Electronics Engineering organization. Northrop Grumman Corporation is a leading global security company whose 120,000 employees provide innovative systems, products, and solutions in aerospace, electronics, information systems, shipbuilding and technical services to government and commercial customers worldwide. Prior to that, he was vice president and general manager of TRW's Space and Technology Division; responsible for spacecraft hardware and software engineering; manufacturing, testing and space vehicle production; as well as chemical and solid-state laser design and development; sensor systems, space and tactical propulsion systems; and research in the physical, chemical and engineering sciences. Dr. Romesser earned a B.S. in physics from Manhattan College and an M.S. and a Ph.D. from the University of Iowa. He is also a graduate of the USC Executive Management Program. Dr. Romesser was elected a fellow of the Directed Energy Professional Society in 2002 and a member of the National Academy of Engineering in 2003.

WILLIAM L. SHELTON retired from the U.S. Air Force in September 2014. His last assignment was as commander, Air Force Space Command, Peterson Air Force Base, Colorado, where he was responsible for organizing, equipping, training, and maintaining mission-ready space and cyberspace forces and capabilities for the North American Aerospace Defense Command, U.S. Strategic Command, and other combatant commands around the world. General Shelton oversaw Air Force network operations; managed a global network of satellite command and control, communications, missile warning and space launch facilities; and was responsible for space system development and acquisition. He led more than 42,000 professionals assigned to 134 locations worldwide. General Shelton entered the Air Force in 1976 as a graduate of the U.S. Air Force Academy. He served in various assignments, including research and development testing, space operations, and staff work. The general has commanded at the squadron, group, wing, and numbered air force levels, and served on the staffs at major command headquarters, Air Force headquarters, and in the Office of the Secretary of Defense. Prior to assuming his final position, General Shelton was the assistant vice chief of staff and the director of Air Staff, U.S. Air Force, the Pentagon, Washington, D.C. He holds an M.S. in astronautical engineering from the U.S. Air Force Institute of Technology, Wright-Patterson AFB, Ohio.

BOB THOMSON is an aerospace consultant and appointed visiting industry director of the Cal Poly CubeSat program. He joined Lockheed Martin in 1981 as an aerodynamics engineer focused on the F-117, U-2, and unmanned vehicle programs. Mr. Thomson assumed a range of progressively more responsible leadership positions, beginning with managing the payload and avionics subsystems for the Dark Star UAV program, evolving to his position as vice president, special programs, leading several multibillion-dollar satellite development programs critical to national security. These satellite development programs spanned the entire procurement life cycle: from the restart of a cold satellite production line, thereby avoiding a national imagery gap, to the capture and start-up of a brand new effort, to completing the integration, test, and subsequent launch of a national asset. His efforts resulted in more than $12 billion dollars of new business for Lockheed Martin. Since his retirement in 2011, Mr. Thomson has been engaged with the dean of engineering at Cal Poly on a variety of special topics. Mr. Thomson is a 1981 graduate of Cal Poly with a B.S. in aeronautical engineering.

DAVID M. VAN WIE is the mission area executive for precision strike at the Johns Hopkins University Applied Physics Laboratory (JHU/APL) with responsibility for the strategic planning, executing, and performance of programs addressing detection and targeting, kinetic engagement, and electronic attack capabilities. Prior to his current assignment, Dr. Van Wie was the chief technologist for the Precision Strike Mission Area, where he focuses on technology development supporting asymmetric mulidomain system concepts for use in anti-access/area-denial environments. Dr. Van Wie holds a research faculty position in the Department of Mechanical Engineering at JHU and has lectured extensively in the Department of Aerospace Engineering at the University of Maryland. He served on committees of the Academies addressing conventional prompt global strike, civil booster systems, and Air Force development planning. Dr. Van Wie also served as a member of the U.S. Air Force Scientific Advisory Board conducting studies on hypersonic systems, small precision weapons, virtual training technologies, future launch vehicles, and munitions for the environment in 2025 and beyond, and he served as the vice chair and chair of the 2010 and 2011 Air Force Research Laboratory Science and Technology Reviews, respectively. Dr. Van Wie is a fellow of the AIAA, an active member of the U.S. science and technology community, and has published extensively in the fields of high-temperature fluid dynamics, plasma aerodynamics, and hypersonic air-breathing propulsion systems.

DEBORAH L. WESTPHAL is managing director of the strategy advisory firm Toffler Associates. Recognized globally for her expertise in strategy, innovation, and organizational transformation, Ms. Westphal helps organizations understand the forces that drive change in their industries and the world, and identifies the best courses of action to create enduring success. Ms. Westphal came to Toffler Associates in 1999 after 13 years as a senior government official in the U.S. Air Force. Her work in the area of technology and advanced concepts for air vehicles, missiles, and space systems has been recognized with numerous awards from the California Air Force Association, a U.S. Air Force Meritorious Civilian Award, an Air Force Association Los Angeles Chapter Civilian of the Year award, and an Air Force Association Medal of Merit. Ms. Westphal has also served on the Army Science Board, the board of directors for the National Defense Industrial Association Greater Los Angeles Chapter, and the board of directors for the Air Force Association, Schriever Chapter 147. Currently, Ms. Westphal serves on the Air Force Studies Board of the Academies. Managing director of Toffler Associates since 2007, she is an acknowledged expert in the aerospace industry and brings a wealth of experience in a wide range of other sectors, including materials, transportation, security, space, hospitality, and telecommunications, as well as U.S. defense, intelligence, and civilian government. Ms. Westphal's success can be traced to her unique combination of education and experience. Holding a B.S. in electrical engineering from the University of New Mexico, she went on to get an MBA from Webster University, and has completed executive education at the Harvard Business School and the Wharton School of Business.

B

Meetings and Speakers

MEETING 1
February 23-24, 2015
Washington, D.C.

Office of the Secretary of Defense (OSD)

Douglas Loverro, Deputy Assistant Secretary of Defense for Space Policy
Karen St. Germain, Deputy Director for Mission Analysis

Intelligence Community

Lawrence Gershwin, National Intelligence Officer for Space and Technical Intelligence
Missile and Space Intelligence Center (DIA)
National Air and Space Intelligence Center (Air Force)
Weapons Intelligence Non-Proliferation and Arms Control (CIA)

MEETING 2
March 19-20, 2015
Washington, D.C.

Office of the Secretary of Defense

Thomas Morgan, Chief, Space Capabilities Division, OUSD(I)

OSD Cost Analysis and Program Evaluation
Steven Miller (SES), Director, Advanced Systems Cost Analysis

Headquarters, U.S. Air Force
Maj Gen Martin Whelan, Director of Space Operations, Headquarters, U.S. Air Force

Headquarters, U.S. Navy
William Flynn (Defense Intelligence Senior Executive Service), Senior Advisor for Space, Chief of Naval Operations, N2N6E

Air Force Space Command
Thomas Walker

U.S. Department of State
HON Frank Rose, Assistant Secretary of State for Arms Control, Verification, and Compliance

National Reconnaissance Office
Stewart Cameron (CIA Senior Intelligence Service), Director, Survivability Assurance Office

MEETING 3
April 7-9, 2015
Washington, D.C.

Inmarsat
Susan Miller, President and Chief Executive Officer
Peter Hadinger, President, U.S. Government Business

Lockheed Martin
Kathy Tobey, Vice President and General Manager
Marc Berkowitz, Strategic Planning Director

Aerospace Corporation
Cathy Steele, Senior Vice President, National Systems Group
Craig Lindsay, Principal Director, Space Control Directorate
Don Lewis, Principal Director, Strategic Awareness and Policy

Northrop Grumman Corporation
Tim Frei, Vice President, Communications Systems

The Boeing Company
Umesh Ketkar, Director, Advanced Space and Intelligence Systems

Intelsat
Rory Welch, Director of Business Development

MIT Lincoln Laboratory
Jay Donnelly, Assistant Division Head, Aerospace Division

Defense Advanced Research Projects Agency
Brad Tousley, Director, Tactical Technology Office

Headquarters, U.S. Air Force
Maj Gen Roger Teague, Director, Space Programs, Office of the Assistant Secretary for Acquisition

U.S. Strategic Command
Evan Hoapili, Associate Director, Capability and Resource Integration

Office of Science and Technology Policy
Travis Blake, Senior Advisor for National Security Space

National Geospatial-Intelligence Agency
John Charles, National GEOINT Officer for Commercial Imagery

Office of the Secretary of Defense
Gil Klinger, Deputy Assistant Secretary of Defense for Space and Intelligence, Office of the Under Secretary of Defense (Acquisition, Technology, and Logistics)

Space Security and Defense Program
Andrew Cox, Director
Russell Partch

MEETING 4
May 7-8, 2015
Washington, D.C.

Office of the Secretary of Defense
James Martin (SES), Director of Defense Intelligence for Intelligence Strategy, Programs, and Resources

MEETING 5
June 1, 2015
Washington, D.C.

Writing meeting

MEETING 6
July 1-2, 2015
Washington, D.C.

Joint Functional Component Command for Space
Lt Gen John W. "Jay" Raymond, Commander 14th Air Force, Air Force Space Command; Commander, Joint Functional Component Command for Space, U.S. Strategic Command

Space and Missile Systems Center
Col Erik C. Bowman, Deputy Director, Space Superiority Systems Directorate, Los Angeles Air Force Base, California

Air Force Research Laboratory
Brandon Arritt, Program Manager, Space Resilience Technologies Air Force Research Laboratory/Space Vehicles Directorate

Sandia National Laboratories
John Rowe, Sandia Fellow
David Cox

U.S. Air Force
Scott Hardiman, Deputy Chief, Space, Aerial and Nuclear Networks, U.S. Air Force

MEETING 7
August 27-28, 2015
Washington, D.C.

Naval Research Laboratory
John Schaub

National Security Agency
Ryan Agee

U.S. Government
Sean R.

MEETING 8
October 8-9, 2015
Washington, D.C.

Writing meeting